Biosensors for Environmental Monitoring

Edited by Toonika Rinken and Kairi Kivirand

Published in London, United Kingdom

IntechOpen

Supporting open minds since 2005

Biosensors for Environmental Monitoring
http://dx.doi.org/10.5772/intechopen.73763
Edited by Toonika Rinken and Kairi Kivirand

Contributors
Xudong Zhang, Sunghoon Jang, William Wang, Fabiano Luiz Naves, Amra Odobašić, Indira Šestan, Sabina Begić, Marcela A. Segundo, Patrícia S. Peixoto, Ana Machado, Hélder P. Oliveira, Adriano A. Bordalo, Kairi Kivirand, Mart Min, Toonika Rinken, Flavio Colmati, Lívia Flório Sgobbi, Ramon Silva Vilela, Giovanna Oliveira Figueiredo, Tatiana Duque Martins, Guilhermina Ferreira Teixeira, Eliana Mossé Alhadeff, Alfredo Jackson Telles Bosco, Caio Fragale Pastusiak, Thais Anjos Correia, Ninoska Isabel Bojorge Ramirez, Patricia Mussali-Galante, Efraín Tovar-Sánchez, Ramón Suarez-Rodríguez, Augusto Ramírez-Trujillo, Leticia Valencia-Cuevas, Isela Hernández-Plata, Geovany Albino De Souza, Shenqi Wang, Umer Farooq, Qiaoli Yang, Muhammad Wajid Ullah

First published in London, United Kingdom, 2019 by IntechOpen
IntechOpen is the global imprint of INTECHOPEN LIMITED, registered in England and Wales, registration number: 11086078, 7th floor, 10 Lower Thames Street, London, EC3R 6AF, United Kingdom
Printed in Croatia

British Library Cataloguing-in-Publication Data
A catalogue record for this book is available from the British Library

Additional hard and PDF copies can be obtained from orders@intechopen.com

Biosensors for Environmental Monitoring
Edited by Toonika Rinken and Kairi Kivirand
p. cm.
Print ISBN 978-1-78923-823-5
Online ISBN 978-1-78923-824-2
eBook (PDF) ISBN 978-1-83962-895-5

Meet the editors

Toonika Rinken is a senior researcher of environmental chemistry and is leading a biosensor development lab at the Institute of Chemistry at the University of Tartu, Estonia. She received her PhD degree in chemistry in 2000 from the same university for her studies on modeling and calibration of biosensors. She has also pursued and completed professional self-improvement in Uppsala (Sweden) and Gröningen (the Netherlands). Dr. Rinken's research focuses on the design and development of biosensing systems for automatic monitoring, and the study of signal rising, modeling, and calibration of biosensor-based analytical systems.

Dr. Kairi Kivirand is a researcher at the University of Tartu, Estonia, and at the Tallinn University of Technology, Estonia. She received a PhD in environmental chemistry in 2011 for her studies of biosensors for biogenic amines. Her research focuses on the purification, identification, and characterization of bioactive molecules and on the design and development of biosensing systems for a wide range of applications. She is an expert in the application of liquid chromatography systems for the extraction and purification of bioactive compounds.

Contents

Preface XI

Chapter 1 1
Introductory Chapter: The Prospective of Biosensing in Environmental Monitoring
by Kairi Kivirand and Toonika Rinken

Chapter 2 7
Normal Boundary Intersection Applied to Controllers in Environmental Controls
by Fabiano Luiz Naves

Chapter 3 19
The Modeling, Design, Fabrication, and Application of Biosensor Based on Electric Cell-Substrate Impedance Sensing (ECIS) Technique in Environmental Monitoring
by Xudong Zhang, William Wang and Sunghoon Jang

Chapter 4 39
Electrochemical Biosensors Containing Pure Enzymes or Crude Extracts as Enzyme Sources for Pesticides and Phenolic Compounds with Pharmacological Property Detection and Quantification
by Flavio Colmati, Lívia Flório Sgobbi, Guilhermina Ferreira Teixeira, Ramon Silva Vilela, Tatiana Duque Martins and Giovanna Oliveira Figueiredo

Chapter 5 53
Challenges and Applications of Impedance-Based Biosensors in Water Analysis
by Kairi Kivirand, Mart Min and Toonika Rinken

Chapter 6 67
Dynamic Luminescent Biosensors Based on Peptides for Oxygen Determination
by Geovany Albino de Souza, Tatiana Duque Martins and Flavio Colmati

Chapter 7 81
Paper-Based Biosensors for Analysis of Water
by Patrícia S. Peixoto, Ana Machado, Hélder P. Oliveira, Adriano A. Bordalo and Marcela A. Segundo

Chapter 8 97
The Use of Biosensors for Biomonitoring Environmental Metal Pollution
by Efraín Tovar-Sánchez, Ramón Suarez-Rodríguez, Augusto Ramírez-Trujillo, Leticia Valencia-Cuevas, Isela Hernández-Plata and Patricia Mussali-Galante

Chapter 9 121
Development of an Ethanol Biosensor Based on Silver Nanoparticles/Polyaniline/Graphite/Epoxy Composite for Friendly Analytical Application
by Eliana Mossé Alhadeff, Alfredo Jackson Telles Bosco, Caio Fragale Pastusiak, Thais Anjos Correia and Ninoska Isabel Bojorge Ramirez

Chapter 10 139
Biosensors for Determination of Heavy Metals in Waters
by Amra Odobašić, Indira Šestan and Sabina Begić

Chapter 11 157
Principle and Development of Phage-Based Biosensors
by Umer Farooq, Qiaoli Yang, Muhammad Wajid Ullah and Shenqi Wang

Chapter 12 175
Applications of Phage-Based Biosensors in the Diagnosis of Infectious Diseases, Food Safety, and Environmental Monitoring
by Umer Farooq, Muhammad Wajid Ullah, Qiaoli Yang and Shenqi Wang

Preface

Expanding industrial production along with the world's growing population, which increases by more than 200,000 individuals per day, puts enormous pressure on the environment. Residues of manmade chemical compounds are everywhere; for example, recent reports indicate that hundreds of organohalogenide contaminants have been found in polar bear blood serum.

There is an urgent need for continuous monitoring of numerous macro and micropollutants and their metabolites in order to cope with the potential threats of extensive exploitation of natural resources and the increasing number of synthetic compounds around. This monitoring is necessary for improving quality of life and for dealing with rising living standards. Currently the contaminants of greatest concern are various micropollutants, such as heavy metals and pesticides used in agricultural production as well as residues of all kinds of pharmaceuticals. A major global threat to public health is "microbiological contamination," that is, antibiotic-resistant bacteria and fungi resulting from uncontrolled use of antibiotics.

Despite the growing need for quick and robust analytical information, we still lack methods for online, onsite automatic detection of contaminants allowing timely and effective management of abnormalities. Traditional analytical methods require sample collection and complex lab studies, taking several hours or even days to provide results. The development of new technologies, in particular various biosensing methods, potentially may allow for real-time, automated analyses in natural conditions with no prior treatment of samples. Although biosensor-based technologies require further efforts to become a serious alternative to standard methods, the number of solutions for the detection of minute residual concentrations of known and potential pollutants along with their metabolites is growing.

This book provides an overview of some of the latest trends in the development of biosensors for their application in environmental monitoring. It discusses different biosensor construction and signal-detection principles along with various sensing platforms and bio-recognition elements. In addition, it proposes algorithms to cope with the variability of analytical conditions and instability of environmental processes. Along with information on artificially constructed sensing systems, the book includes chapters dealing with natural "biosensor" organisms used for the assessment of detrimental effects of pollutants.

In addition, this book pays special attention to the analytical performance of biosensors, including their selectivity, sensitivity, and analysis time as well as their potential application for automated online analyses.

Toonika Rinken and Kairi Kivirand
University of Tartu, Estonia

Chapter 1

Introductory Chapter: The Prospective of Biosensing in Environmental Monitoring

Kairi Kivirand and Toonika Rinken

1. Introduction

In recent years, an increasing number of actions for environmental monitoring have become more and more topical. Environmental monitoring is necessary to protect the environment from pollutants and minimize the impact of unfavorable components and processes. As the world's population continues to increase, so does the amount of pollutants that are released into the environment.

Environmental pollution is defined as "the contamination of the physical and biological components of the earth/atmosphere system to such an extent that normal environmental processes are adversely affected" [1]. The substances that cause pollution are categorized as pollutants, which are commonly classified according to their chemical structure (organic and inorganic compounds), their mode of action (endocrine effect or toxicity), their source (natural or manmade) or by their amount (micro and macro pollutants) [2]. A pollutant can be any chemical or geochemical substance, biological organism or physical substance, which has been released into the environment by man and has harmful, unpleasant or inconvenient effect [3]. Depending on the nature of a pollutant, pollution is classified as air, water, soil or land, noise, radioactive and thermal pollution [3].

The release of pollutants (e.g., heavy metals, pesticides, drugs and pharmaceuticals) to the environment is a worldwide problem and there is a growing need to combat with uncontrolled pollution. For example, the global environmental monitoring plan (GEMP) for persistent organic pollutants (POPs), prefiguring a major concern, has become an important component of the evaluation of effectiveness of Stockholm Convention [4]. It provides an organizational framework for the collection of comparable monitoring data on the presence of POPs in order to detect changes in their concentrations [5]. Most pollutants released to the environment are undetectable, until their effects make it impossible to ignore them and we have to deal already with the consequences. Therefore, it is necessary to detect pollution as soon as possible.

2. Traditional methods for environmental analysis

Different types of methods and techniques are used for environmental analysis. Traditional methods used for the detection of molecular pollutants are mostly based on chromatographic techniques (gas chromatography or ultra-high performance liquid chromatography coupled with mass spectrometry, thin-layer chromatography) and spectrophotometry. Chromatographic tools are sensitive and reliable but also time-consuming because they usually need sample pretreatment; the equipment is expensive and requires qualified personnel to perform the analysis. The biggest drawback of the abovementioned methods is the fact that due to long analytical procedures, their application for operative in situ measurements in cases when timely information is crucial is not possible. For example, pollutant concentrations in watercourses are dynamic and change in water flows. With weekly or even monthly sampling and analyzing, it is extremely unlikely that the maximum or the real concentration levels for a certain period can be determined. As a result, we see elevated levels of pesticides or nitrates in areas of intensive farming, even in groundwater, or increased lead levels in areas where it has been used in plumbing. In addition, thin-layer chromatography (TLC) has been often used for testing soils and groundwater for pollutants like pesticides, herbicides or fungicides. It is an effective and low-cost method and many samples can be analyzed simultaneously. However, TLC is applicable only for nonvolatile compounds; there are limitations in resolution capability and the absence of fully automated system [6].

The gold standard for the detection of microbiological pollutants is cultivation; however, DNA-based molecular diagnostics nowadays is becoming more and more popular. Microbiological cultivation is simple and inexpensive. However, there are some disadvantages: these methods rely on the growth of the target microorganisms in one or more nutrient media, the detection of growth is carried out by visual assessment and the confirmation of the presence of a particular pathogen usually involves a combination of biochemical and serological tests [7, 8]. In addition, the interpretation of the results can be subjective, and for some bacteria, the total test time can be as long as several days [8, 9]. For example, there is a drastic increase in pathogen concentrations, and the infection risks due to late results of potentially contaminated drinking water are very high [10]. With DNA-based molecular diagnostic methods like polymerase chain reaction (PCR), it is possible to identify specific bacterial strains, but this method still require several hours to obtain results and sometimes it fails to discriminate between related species and intragenomic heterogeneity [11].

3. Biosensing in environmental monitoring

Therefore, development of rapid real-time and reliable detection methods is essential. Biosensors are an alternative to traditional methods. Biosensors can act as pressure sensors, microphones, optical sensors, microfluidics, temperature and gas sensors [12]. In recent years, biosensors have also been developed to detect and recognize genetically modified microorganisms (GMOs) [13], which have generated heated debates, especially in the European countries (EU), about the safety of food and the potential impact to the environment [14]. Furthermore, biosensors can offer a strong potential for better understanding and investigating of the environment, including the fate and transport of contaminants. The number of opportunities to join science and new technology into biosensing systems is almost overwhelming. One of the first environmental biosensors was initially developed for nerve gas detection for the military in the late 1970s and modified for the detection of pesticides (organophosphorus and carbamate) in the environment and was based on the inhibition of the enzyme acetylcholinesterase [15]. Over the years, new biosensors have been developed for environmental monitoring. For example, biosensors for the detection of heavy metals like zinc, cobalt, cadmium, lead, etc. [16–22] have been developed. In addition, biosensors for the detection of phenolic compounds [23–26], pesticides [27–30], pathogens [9, 31–35] and drug residues [36, 37] have been developed.

Sensitivity, specificity, reliability, cost-efficiency and the possibility of on-line use are crucial factors for the design and construction of a biosensor for environmental monitoring. Functional bio-recognition elements are the key components, which define the affinity (low detection limit), specificity (low interference), dynamic range, response time and lifetime of the biosensing system.

Although most of the developed biosensor-based methods are not able to compete with traditional methods in terms of precision or reproducibility, they allow continuous on-site and real-time monitoring of a contaminated area and provide timely information about potential pollution.

Currently, only a few commercial biosensors are available for environmental monitoring [38]. Up to date, the biochemical oxygen demand (BOD) sensor, which was invented in Japan in the late 1970s, has commercially been the most successful biosensor for environmental applications.

Author details

Kairi Kivirand[1,2]* and Toonika Rinken[1]

1 Institute of Chemistry, University of Tartu, Tartu, Estonia

2 Thomas Johann Seebeck Department of Electronics, Tallinn Technical University, Tallinn, Estonia

*Address all correspondence to: kairi.kivirand@ut.ee

References

[1] Muralikrishna IV, Manickam V. Environmental Management: Science and Engineering for Industry. UK: Butterworth-Heinemann; 2017. p. 664. DOI: 10.1016/B978-0-12-811989-1.00001-4

[2] Rodriguez-Mozaz S, Lopez De Alda MJ, Barceló D. Biosensors as useful tools for environmental analysis and monitoring. Analytical and Bioanalytical Chemistry. 2006;**386**:1025-1041. DOI: 10.1007/s00216-006-0574-3

[3] Rai PK. Biomagnetic Monitoring of Particulate Matter: In the Indo-Burma Hotspot Region. Elsevier; 2001. p. 216. DOI: 10.1016/B978-0-12-805135-1.00001-9

[4] Global Monitoring Plan [Internet]. 2008. Available from: http://www.pops.int/Implementation/GlobalMonitoringPlan/Overview/tabid/83/Default.aspx

[5] Stockholm Convention [Internet]. Available from: http://chm.pops.int/

[6] Cai L. Thin layer chromatography. Current Protocols Essential Laboratory Techniques. 2014;**8**:6.3.1-6.3.18. DOI: 10.1002/9780470089941.et0603s08

[7] Brooks JL, Mirhabibollahi B, Kroll R. Experimental enzyme-linked amperometric immunosensors for the detection of salmonellas in foods. Journal of Applied Bacteriology. 1992;**73**:189-196. DOI: 10.1111/j.1365-2672.1992.tb02977.x

[8] McCarthy J. Immunological techniques: ELISA. In: Woodhead Publishing Series in Food Science, Technology and Nutrition, Detecting Pathogens in Food. England: Woodhead Publishing; 2003. pp. 241-258. DOI: 10.1533/9781855737044.2.241

[9] Juronen D, Kuusk A, Kivirand K, Rinken A, Rinken T. Immunosensing system for rapid multiplex detection of mastitis-causing pathogens in milk. Talanta. 2018;**178**:949-954. DOI: 10.1016/j.talanta.2017.10.043

[10] Cabral JPS. Water microbiology. Bacterial pathogens and water. International Journal of Environmental Research and Public Health. 2010;**7**(10):3657-3703. DOI: 10.3390/ijerph7103657

[11] Talkington DF. PCR detection of microbial pathogens. Emerging Infectious Diseases. New York: Humana Press; 2013;**19**(8):1353. DOI: 10.3201/eid1908.130528

[12] Nigam VK, Shukla P. Enzyme based biosensors for detection of environmental pollutants—A review. Journal of Microbiology and Biotechnology. 2015;**25**(11):1773-1781. DOI: 10.4014/jmb.1504.04010

[13] D'Souza SF. Microbial biosensors. Biosensors and Bioelectronics. 2001;**16**(6):337-353. DOI: 10.1016/s0956-5663(01)00125-7

[14] López MSP, Manzanares-Palenzuela CL, López-Ruiz B. Biosensors for GMO testing: Nearly 25 years of research. Critical Reviews in Analytical Chemistry. 2018;**48**(5):391-405. DOI: 10.1080/10408347.2018.1442708

[15] Singh RL. Principles and Applications of Environmental Biotechnology for a Sustainable Future. Singapore: Springer; 2015. pp. 1-487. DOI: 10.1007/978-981-10-1866-4

[16] Ma H, Sang-Bing T. Design of research on performance of a new iridium coordination compound for the detection of Hg^{2+}. International Journal of Environmental Research and Public

Health. 2017;**14**(10):1232. DOI: 10.3390/ijerph14101232

[17] Wong ELS, Crow E, Gooding JJ. The electrochemical detection of cadmium using surface-immobilized DNA. Electrochemistry Communications. 2007;**9**(4):845-849. DOI: 10.1016/j.elecom.2006.11.018

[18] Pal P, Bhattacharyay D, Mukhopadhyay A, Sarkar P. The detection of mercury, cadium, and arsenic by the deactivation of urease on rhodinized carbon. Environmental Engineering Science. 2009;**26**(1):25-32. DOI: 10.1089/ees.2007.0148

[19] Knecht MR, Sethi M. Bio-inspired colorimetric detection of Hg^{2+} and Pb^{2+} heavy metal ions using Au nanoparticles. Analytical and Bioanalytical Chemistry. 2009;**394**:33-46. DOI: 10.1007/s00216-008-2594-7

[20] Ilangovan R, Daniel D, Krastanov A, Zachariah C, Elizabeth R. Enzyme based biosensor for heavy metal ions determination. Biotechnology and Biotechnological Equipment. 2006;**20**(1):184-189. DOI: 10.1080/13102818.2006.10817330

[21] Turdean GL. Design and development of biosensors for the detection of heavy metal toxicity. International Journal of Electrochemistry. 2011;**2011**:1-15. DOI: 10.4061/2011/343125

[22] Chen J, Rosen BP. Biosensors for inorganic and organic arsenicals. Biosensors. 2014;**4**:494-512. DOI: 10.3390/bios4040494

[23] Karim F, Fakhruddin ANM. Recent advances in the development of biosensor for phenol: A review. Reviews in Environmental Science and Biotechnology. 2012;**11**:261-274. DOI: 10.1007/s11157-012-9268-9

[24] Ozoner SK, Erhan E, Yilmaz F. Enzyme based phenol biosensors. In: Somerset V, editor. Environmental Biosensors. UK: IntechOpen; 2011. DOI: 10.5772/16998

[25] Osma JF, Stoytcheva M. Biosensors: Recent Advances and Mathematical Challenges. Barcelona, Spain: OmniaScience; 2014. DOI: 10.1016/j.snb.2012.07.026

[26] Wang Y, Zhai F, Hasebe Y, Jia H, Zhang Z. A highly sensitive electrochemical biosensor for phenol derivatives using a graphene oxide-modified tyrosinase electrode. Bioelectrochemistry. 2018;**122**:174-182. DOI: 10.1016/j.bioelechem.2018.04.003

[27] Verma N, Bhardwaj A. Biosensor technology for pesticides—A review. Applied Biochemistry and Biotechnology. 2015;**175**:3093-2119. DOI: 10.1007/s12010-015-1489-2

[28] Bucur B, Munteanu FD, Marty JL, Vasilescu A. Advances in enzyme-based biosensors for pesticide detection. Biosensors. 2018;**8**(2):1-28. DOI: 10.3390/bios8020027

[29] Cortina-Puig M, Istamboulie G, Noguer T, Marty JL. Analysis of pesticide mixtures using intelligent biosensors. In: Somerset VS, editor. Intelligent and Biosensors. UK: IntechOpen; 2010. DOI: 10.5772/7154

[30] Vargas-Bernal R, Rodríguez-Miranda E, Herrera-Pérez G. In: Soundararajan RP, editor. Evolution and Expectations of Enzymatic Biosensors for Pesticides. Pesticides—Advances in Chemical and Botanical Pesticides. UK: IntechOpen; 2012. DOI: 10.5772/46227

[31] Oluwaseun AC, Phazang P, Sarin NB. In: Rinken T, Kivirand K, editors. Biosensors: A Fast-Growing Technology for Pathogen Detection in Agriculture and Food Sector, Biosensing

Technologies for the Detection of Pathogens - a Prospective Way for Rapid Analysis. UK: IntechOpen; 2018. DOI: 10.5772/intechopen.74668

[32] Justino CIL, Duarte AC, Rocha-Santos TAP. Recent progress in biosensors for environmental monitoring: A review. Sensors. 2017;**17**(12):1-25. DOI: 10.3390/s17122918

[33] Peedel D, Rinken T. Rapid biosensing of *Staphylococcus aureus* bacteria in milk. Analytical Methods. 2014;**6**:2642-2647. DOI: 10.1039/c3ay42036a

[34] Viirlaid E, Riiberg R, Mäeorg U, Rinken T. Glyphosate attachment on aminoactivated carriers for sample stabilization and concentration. Agronomy Research. 2015;**13**(4):1152-1159. Available from: http://agronomy.emu.ee/vol134/13_4_29_B5.pdf

[35] Kuusk E, Rinken T. Transient phase calibration of tyrosinase-based carbaryl biosensor. Enzyme and Microbial Technology. 2004;**34**(7):657-661. DOI: 10.1016/j.enzmictec.2004.03.004

[36] Kagan M, Printsmann G, Kivirand K, Rinken T. Determination of penicillins in milk by a dual-optrode biosensor. Analytical Letters. 2017;**50**(5):819-828. DOI: 10.1080/00032719.2016.1202957

[37] Barceló D, Hansen PD. Biosensors for the Environmental Monitoring of Aquatic Systems. Berlin Heidelberg: Springer-Verlag; 2009. DOI: 10.1007/978-3-540-36253-1

[38] Burcu E, Mustafa B, Sezgintürk K. Review applications of commercial biosensors in clinical, food, environmental, and biothreat/biowarfare analyses. Analytical Biochemistry. 2015;**478**:107-120. DOI: 10.1016/j.ab.2015.03.011

Chapter 2

Normal Boundary Intersection Applied to Controllers in Environmental Controls

Fabiano Luiz Naves

Abstract

Generally, the controllers currently used and implemented in the environmental field have certain set point values, which are pre-calibrated according to a specific process characteristic. However, instability in environmental processes is a difficult variable to fix. Thus, the use of numerous set points for specific process conditions may be a way of controlling instability. One way to obtain numerous setups within a working region is to use optimization algorithms for the construction of the Pareto frontier, each point of the boundary being represented by a different and at the same time optimum setup of operation. In this context, the construction of a Pareto frontier for a multiobjective and multivariate problem, established from an environmental problem, can be a way of getting around the problem of process instability. This chapter has a main objective to demonstrate the possibility of using the algorithm Normal Boundary Intersection (NBI), originally enunciated by Karna, as a precursor for the construction of the Pareto frontier, as well as the possibility of implementing the generated function for implementation in programmable logic systems.

Keywords: NBI, controllers, multivariate optimization, environmental, biosensors

1. Introduction

Comprehensively, much of the real industrial processes make use of several input variables (factors) at levels often unpredictable due to the instability displayed during operation in transient regime. The actual processes are very difficult to control, especially when it comes to numerous responses to be controlled.

Figure 1 shows in an illustrative way a real process where other factors that could directly influence the responses and interactions, called noise, were not considered. These noises can be related from the events of the environment where the process occurs, such as variations in the temperature of the medium, or events related to errors occurred by the operators. The use of complete second-order models to model processes should be restricted to only a certain interval specified by the levels presented for each of the factors analyzed. In the context of environmental processes, such as effluent decontamination in a treatment plant, the waste disposal parameters are defined according to country-specific standards and must be strictly followed. By using the effluent treatment plant as an example, it is practically impossible to maintain the constant input parameters such as the

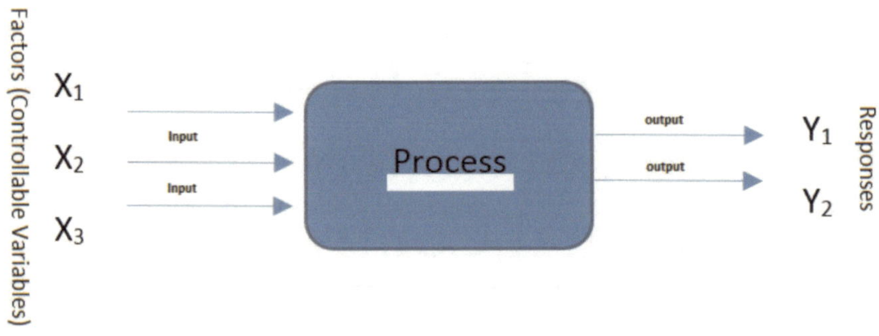

Figure 1.
General process diagram.

incoming organic load, heavy metals, and turbidity, among others. In order to keep the process running at steady state, with possible variations of input, it is necessary that the levels of the controllable factors be adequate in order to keep the responses at the exit within the pre-established parameters. This adjustment of levels can be achieved through sensors connected to programmable logic controllers, which usually operate through a set point.

These types of equipment are microprocessor computers that perform the function of control through specific software. One of the major problems encountered in using this device can be attributed to numerous, generally correlated, input conditions that may occur throughout the operation. Even with controller performance due to the set point, it no longer considers the possibility of interactions between these input parameters, which may compromise the permanence of the steady state. When the noise source is not discovered for later quantification, instability in the process can lead to desired responses outside of the predefined standards, leading to losses, associated cost, and environmental damage. Therefore, the controllers currently applicable cannot consider this instability generated by the noise industrially.

In the environmental area, due to the large number of parameters that must be monitored and pre-established as waste disposal control standards in receiving bodies, it is very common to maintain a certain operation for numerous responses. Thus, it is fundamental that the process can be previously known, modeled, and later optimized through algorithms already fomented by the literature, allowing the implementation of robust multiobjective optimization from the polynomial that describes all the responses, factors, and levels of the process in detail.

The concept of multiobjective robust optimization can be described as the set of nonlinear constrained programming (NLP) methods and algorithms that are intended to simultaneously optimize the mean and variance of multiple process characteristics that are in a way correlated output quantities that are reasonably well modeled by complete quadratic models. However, in effluent treatment processes that have multiple output characteristics are generally correlated.

Any process can be defined through a quadratic polynomial, if it is properly constrained within certain predefined intervals. The original concept of "robust" process was introduced by Genichi Taguchi in 1980 [1]. To this concept we can associate the original idea of RPD (robust parameter design), applied to generic processes. The more "robust" the details of the process are known, the more accurately it can be modeled and optimized. Therefore, there are several situations in which the multiple means of responses must be optimized and the multiple variances associated with each of the responses individually, minimized. This routine can be performed in order to reduce the interference attributed to the noise and to maintain a more stable process. As already mentioned, independently of the innumerable responses to be analyzed to a process, they are easily analyzed individually, even knowing the existence of a high associated positive correlation. Thus, when the responses have a very high correlation, mainly positive, very common in processes that involve chemical reactions, it becomes impracticable to perform the modeling of the multiple objective functions in an independent way, leading to the wrong responses.

In multiobjective optimization problems, the assignment of convex combinations of weights to the multiple responses leads to the agglutination of the objective functions that represent each response through weighted sums, thus generating a Pareto border or surface. Pareto border or surface is therefore a set of optimal values for multiple features obtained from a list of viable optimal points, obviously within a region of viable space. This agglutination of functions can be performed according to some methods: weighted sum and global criterion method (GCM). Both allow the construction of the Pareto border with some constraints attributed to the convexity of the objective function presented in the region of the viable space where the boundary is constructed. When there is a non-convex region in a certain objective function to be analyzed, the Pareto boundary cannot detect optimal points in this region.

Analyzing **Figure 2**, it is possible to verify a Pareto frontier for two responses, where each of the points represents different operating conditions. However, there is a discontinuity indicating no convex region of both functions representing the responses. One way to solve this problem is to use the algorithm Normal Boundary Intersection (NBI) to construct the Pareto frontier. This algorithm is able to determine points along the boundary, even in non-convex regions of space. The NBI algorithm considers two fixed points of the frontier ("best of the best and worst of the worst") known as utopia and nadir respectively. Between these fixed points, all others that make up the border are distributed.

One of the great possibilities in using this algorithm as a transfer function in control processes is precisely the possibility of choosing a number of different process setups, which consequently lead to optimized responses between the utopia and nadir points, which may be the limits of specification of the particular disposal

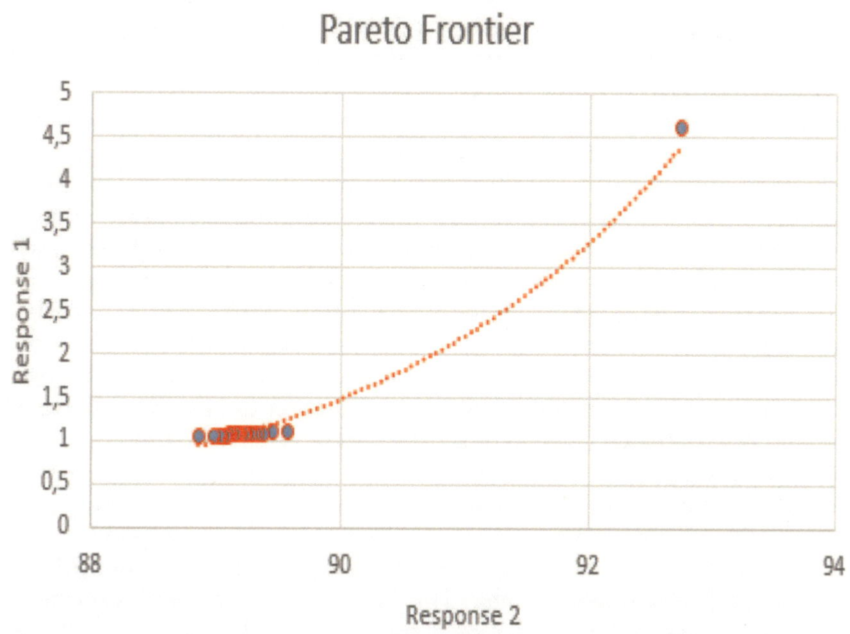

Figure 2.
Pareto frontier showing discontinuity.

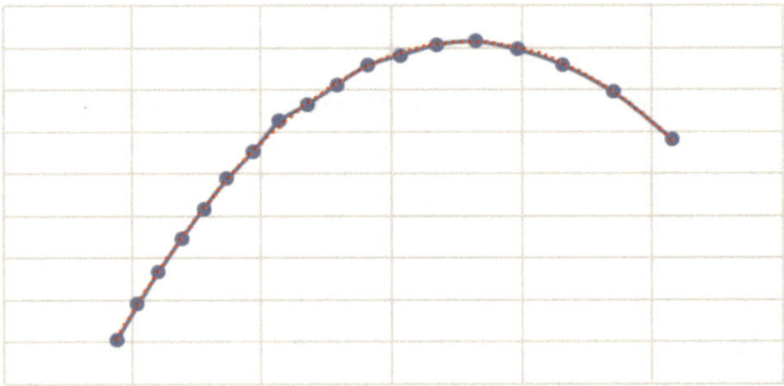

Figure 3.
Pareto border for bi-objective problem.

parameter in an effluent treatment plant, for example. The polynomial (**Figure 3**), which represents the Pareto frontier, can be used as a transfer function in scaling of a possible dynamic process controller for multiobjectives.

2. Description of the process NBI control

In order to facilitate the understanding of the possibility of implementing the NBI algorithm in controllers, let us take, for example, an industrial effluent treatment plant, which operates with a certain constant flow, due to the residence time necessary for part of the organic load to be degraded via bacteria and protozoa in an aerobic process. As a base of the input variables, we will work with initial organic load in terms of biochemical oxygen demand (BOD) and pH. As controllable factors, we will use the air or oxygen flow (aeration) and residence time. As desired responses, we will use as an illustration the removal of the organic load in terms of biochemical oxygen demand (BOD) and chemical oxygen demand (COD).

Modeling a typical problem processes, we could write that both responses have a direct relationship with the two factors presented as X_1 and X_2. However, keeping the process steady relative to the inputs becomes virtually impossible. By establishing, the two responses used, as an illustration of the application of the method, is it feasible to predict the aeration rate and residence time required. Certainly, the answer would be positive, if the entries were kept constant. However, if this standardization is not possible, how can we keep the responses within desirable patterns? Imagine in a situation of actual biological treatment process, where some changes can lead to periodic changes in the conditions of entry. For example, an increase in the rainfall rate may lead to the dilution of the organic matter present in the tributary and consequently the decrease of the initial BOD. The decrease of the initial BOD requires a lower concentration of dissolved oxygen so that bacteria and protozoa can decompose the organic matter in order to meet the exit standards, which would lead to the conclusion of shorter residence times required. There is a relationship as presented that can be considered a certainty. However, what is the relationship between them? What would be the best condition, to decrease aeration or increase residence time? These responses can only be met if we have this problem modeled. When working with models, we can easily predict the relationship of each of the factors to the expected response. This fact helps us reduce process costs and increase effectiveness in the targeted response. Through the use of models created from response surfaces, which have quadratic models, it is easily possible to determine local or global minimum or maximum points.

2.1 Stochastic response surface models

The response surface methodology (MSR) is a collection of mathematical and statistical techniques that allows modeling, analyzing, and optimizing problems whose response variables are influenced by many variables [2]. As mentioned earlier, there is great difficulty in knowing the behavior of independent and dependent variables in a process. Thus, the response surface allows the real approximation of the process from a quadratic model. The development through a Taylor polynomial, truncated in the quadratic term, takes what we call a second-order response surface:

$$Y(\mathbf{x}) = \beta_0 + \sum_{i=1}^{k} \beta_i x_i + \sum_{i=1}^{k} \beta_{ii} x_i^2 + \sum_{i<j} \sum \beta_{ij} x_i x_j + \varepsilon \tag{1}$$

where β represents the coefficients of the model, k is the number of independent variables considered in the study, and ε is the error term.

The fact of using the response surface in a region close to high curvature of the model, presented according to local or global maxima or minima, according to convexity, does not effectively determine the best points or operation setups. However, what can be verified is a region of space that, depending on the levels of each of the independent variables, leads to better responses.

From the color gradient shown in **Figure 4**, it is possible to verify regions, delimited through the Cartesian axes representing the levels of each of the factors studied, leading to better responses. Thus, the construction of models through the surface response method becomes paramount for the application of later optimization algorithms. Among several optimization algorithms, the Normal Boundary Intersection (NBI) [3] has been used in several researches, in several different fields.

2.2 NBI algorithm

The NBI algorithm is developed in terms of an array that we call the payoff matrix Φ , which represents the optimal values of the multiple objective functions

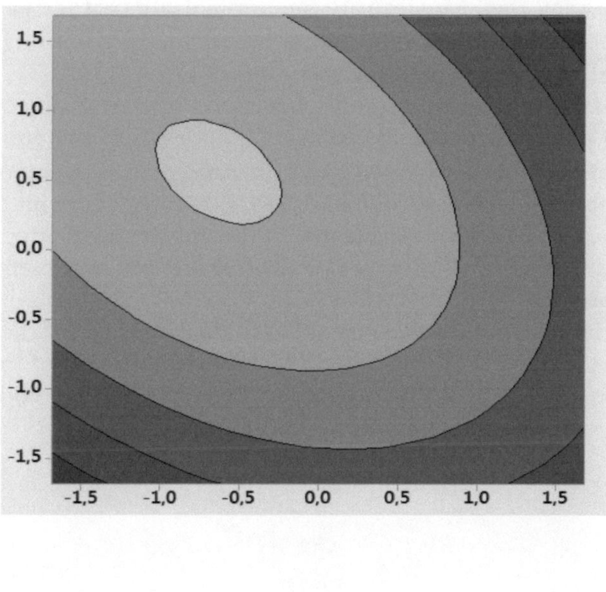

Figure 4.
Counter graphic.

minimized individually. The solution vector that minimizes the i-th objective function individually $f_i(x)$ is represented by x_i^* so that the minimum value of $f_i(x)$ at this point is $f_i^*(x_i^*)$. When replacing the individual optimum point x_i^* obtained in the optimization of objective function in the other functions, we have $f_i(x_i^*)$ which is therefore a nonoptimal value of this function. By repeating this algorithm for all functions, we can represent the payoff matrix as

$$\Phi = \begin{bmatrix} f_1^*(x_1^*) & \cdots & f_1(x_i^*) & \cdots & f_1(x_m^*) \\ \vdots & \ddots & & & \vdots \\ f_i(x_1^*) & \cdots & f_i^*(x_i^*) & \cdots & f_i^*(x_m^*) \\ \vdots & & & \ddots & \vdots \\ f_m(x_1^*) & \cdots & f_m(x_i^*) & \cdots & f_m^*(x_m^*) \end{bmatrix} \quad (2)$$

Each line of Φ is composed of minimum and maximum values of $f_i(x)$. In the NBI method, these values can be used to normalize the objective functions, especially when they are represented by different scales or units. In a similar way, writing the set of individual optimums in a vector, we have

$$f^U = [f_1^*(x_1^*) \; ..., \; f_i^*(x_i^*) \; ..., \; f_m^*(x_m^*)]^T \quad (3)$$

This vector is called utopia point. In the same way, by grouping the maximum (nonoptimal) values of each objective function, we have

$$f^N = [f_1^N \; ..., \; f_i^N \; ..., \; f_m^N]^T \quad (4)$$

This vector is called nadir points.

Using these two sets of extreme points, the normalization of the objective functions can be obtained as

$$\bar{f}(x) = \frac{f_i(x) - f_i^U}{f_i^N - f_i^U}, i = 1, ..., m \quad (5)$$

This normalization therefore leads to the normalization of the payoff matrix, $\overline{\Phi}$. The convex combinations of each line of the payoff matrix, $\overline{\Phi}$, form the "convex hull of individual minima" (CHIM) or the utopia line.

Figure 5 illustrates the main elements associated with multiobjective optimization. The anchor points represent the individual solutions of two functions. Points a and b are calculated from the stepped payoff matrix, $\overline{\Phi}\, w_i$. Considering a set of convex values for the weights, w, one has to $\Phi\, w_i$ represent a point on the utopia line, making $\hat{n}$ denote a unit vector normal to the line at point's utopia $\Phi\, w_i$ in the direction of origin; at the time, $\Phi\, w + D\,\hat{n}$ with $D \in R$ will represent the set of points in that normal.

The point of intersection of this normal with the boundary of the viable region that is closest to the origin will correspond to the maximization of the distance between the utopia line and the Pareto border. Thus, the NBI method can be written as a constrained nonlinear programming problem such that

$$\begin{aligned} &\underset{(\mathbf{x},t)}{\text{Max}} && D \\ &\textit{subject to}: && \overline{\Phi} w + D\hat{n} = \overline{F}(\mathbf{x}) \\ & && \mathbf{x} \in \Omega \end{aligned} \quad (6)$$

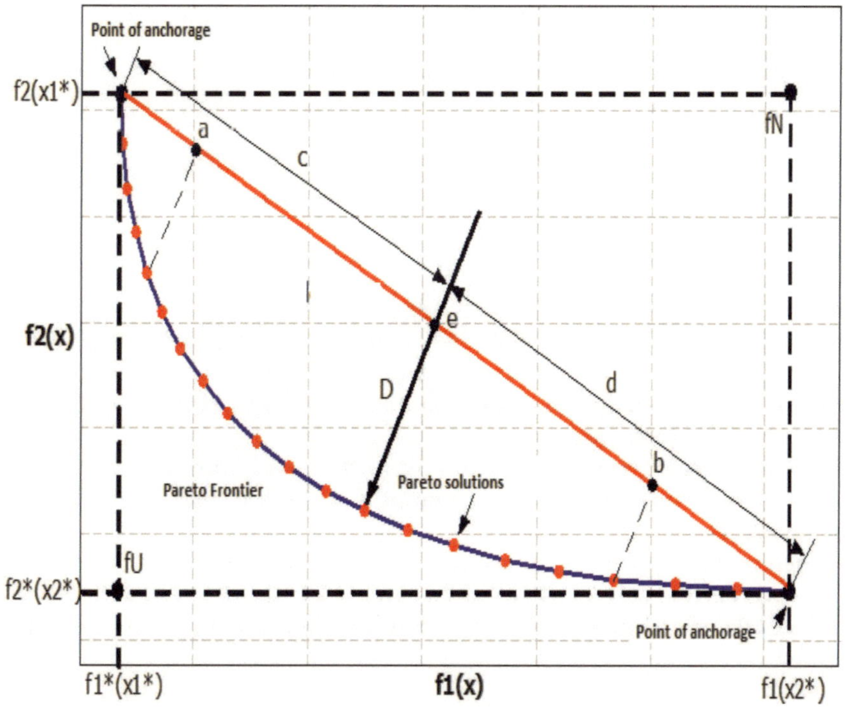

Figure 5.
Normal to intersect method (NBI).

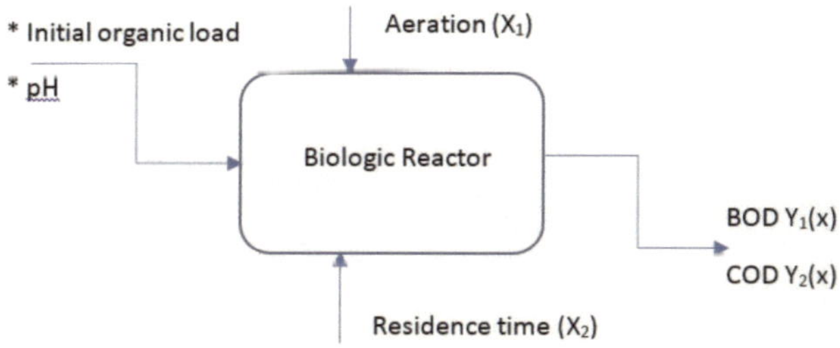

Figure 6.
General scheme of the process.

2.3 Implementation of the NBI control system

For the process described as an example, there are two controllable factors represented by the aeration rate (x_1) and residence time (x_2). However, according to **Figure 6**, there are also two input variables that cannot be measured, mainly due to the instability of a biological treatment plant, according to initial organic charge z_1 and pH z_2. The first artifice presented will be the transformation of each of these variables into known values, from experiments carried out on a smaller scale.

Thus, we will have the following factors: aeration rate (x_1), residence time (x_2), initial organic load (x_3), and pH (x_4). From a surface of response called central composite design (CCD), it is possible to construct a quadratic model, executing 31 experiments in laboratory scale:

$$Y1_x = \beta_0 + \beta_1 x_1 + \beta_2 x_2 + \beta_3 x_3 + \beta_4 x_4 + \beta_{11} {x_1}^2 + \beta_{22} {x_2}^2 + \beta_{33} {x_3}^2 + \beta_{44} {x_4}^2 + \beta_{12} x_1 x_2 + \beta_{13} x_1 x_3 + \beta_{14} x_1 x_4 + \beta_{23} x_2 x_3 + \beta_{24} x_2 x_4 + \beta_{34} x_3 x_4 + \varepsilon \quad (7)$$

$$Y2_x = \beta_0 + \beta_1 x_1 + \beta_2 x_2 + \beta_3 x_3 + \beta_4 x_4 + \beta_{11} {x_1}^2 + \beta_{22} {x_2}^2 + \beta_{33} {x_3}^2 + \beta_{44} {x_4}^2 + \beta_{12} x_1 x_2 + \beta_{13} x_1 x_3 + \beta_{14} x_1 x_4 + \beta_{23} x_2 x_3 + \beta_{24} x_2 x_4 + \beta_{34} x_3 x_4 + \varepsilon \quad (8)$$

Each of the coefficients presented in the two equations, represented by β_i, β_{ii}, and β_{ij}, is determined by the ordinary least square (OLS) regression algorithm where x_1, x_2, x_3, and x_4 are the factors already stated. With the models presented, it is possible to propose an optimization of both responses from the NBI algorithm for the four factors (**Figure** 7).

The Pareto frontier constructed from the optimum of both responses can now, from each of the setups assigned to each point, serve as the basis for implementation in controllers.

For each point referring to the specific response condition, a different setup is considered. For the chosen point 1 according to **Figure 8**, there is a BOD of 33.2 and a COD of 67, and under these conditions, we have the levels of each of the factors:

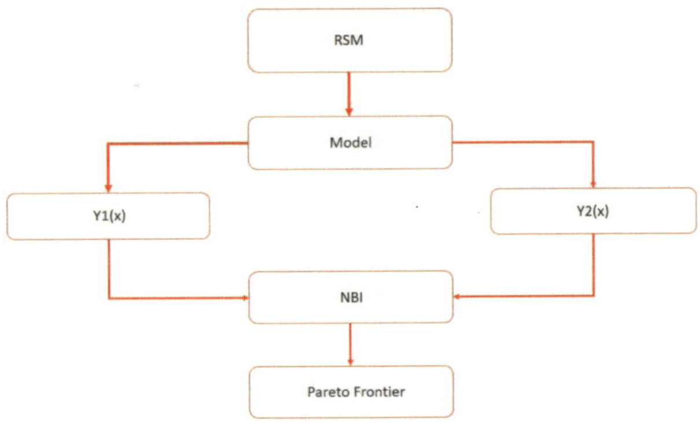

Figure 7.
Modeling and optimization flowchart.

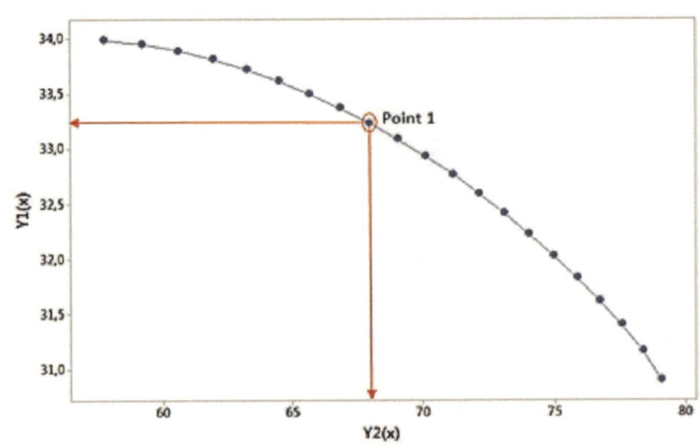

Figure 8.
Pareto frontier with sample choice point.

$$\begin{cases} \text{Aeration rate} = x_1 \\ \text{Residence time} = x_2 \\ \text{Initial organic load} = x_3 \\ \text{pH} = x_4 \end{cases} \tag{9}$$

In the conditions of this chosen point, replacing in (Eqs. (6) and (7)) the response surface, we have two quadratic equations, one referring to Y_1 (x) and Y_2 (x).

The implementation of the transfer function in the control will be done according to **Figure 9**.

The two responses provided in the example, enter into a multiprocessor system according to pre-established parameters. The multiprocessing system introduces

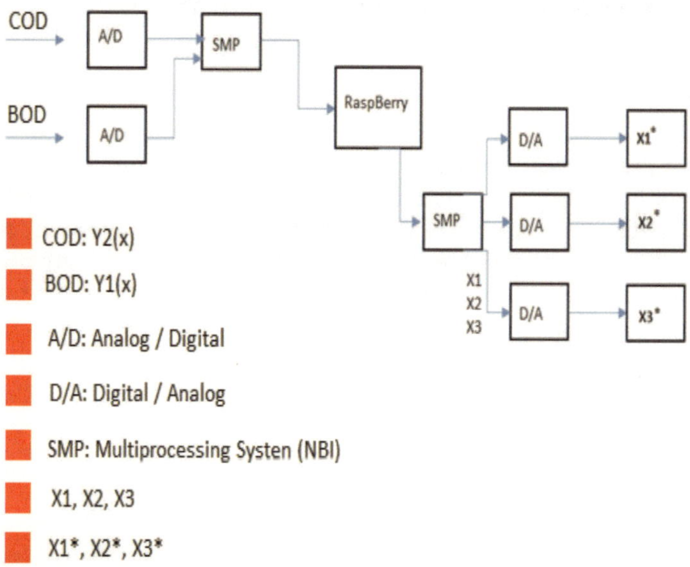

Figure 9.
Proposed arrangement for implementation.

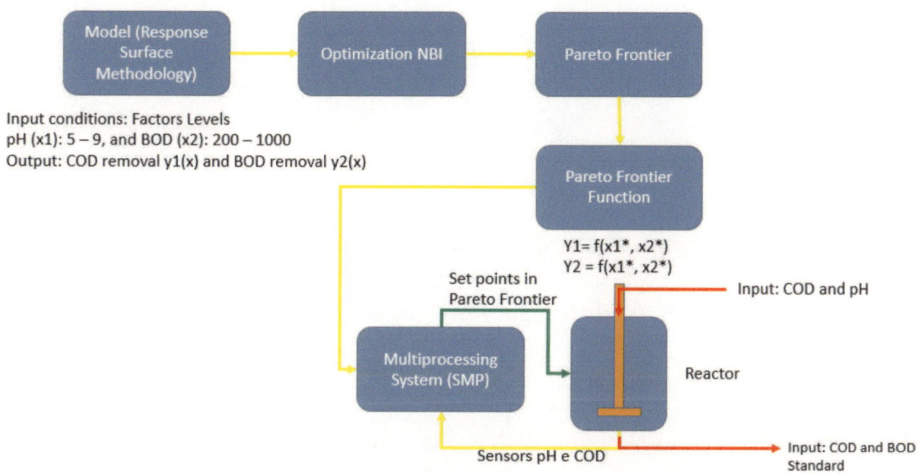

Figure 10.
Flow sheet of implementation algorithm.

polynomials referring to each different setup that consisted of the Pareto frontier, and for each setup, there are specific values of COD and BOD in mgO_2L^{-1}. From these inputs, the factors can be determined in optimized terms, X_1 *, X_2 *, and X_3 *.

An example of implementation for pH 5–9 and BOD values between 200 and 1000 mgL^{-1} follows the flow sheet (**Figure 10**).

As already mentioned, one of the advantages of the method is the correction of the input parameters belonging to the Pareto frontier, consisting of innumerable set points within an optimal solution space.

3. Conclusions

Although it has not yet been implemented in controllers, the use of algorithms such as NBI can facilitate the operation of this equipment, as well as lower costs of implementation and operation of environmental systems.

Conflict of interest

The author certify that they have no affiliations with or involvement in any organization or entity with any financial interest (such as honoraria; educational grants; participation in speakers' bureaus; membership, employment, consultancies, stock ownership, or other equity interest; and expert testimony or patent-licensing arrangements) or nonfinancial interest (such as personal or professional relationships, affiliations, knowledge, or beliefs) in the subject matter or materials discussed in this chapter book.

Author details

Fabiano Luiz Naves
University Federal of São João del Rei—UFSJ, Minas Gerais, Brazil

*Address all correspondence to: fabianonavesengenheiro@ufsj.edu.br

References

[1] Karna SK. An overview on Taguchi method. Society for Industrial and Applied Mathematics. 2016;**1**:10-17

[2] Montgomery DC. Design and Analysis of Experiments Eighth Edition. 2013

[3] Das I, Dennis JE. Normal-boundary intersection: A new method for generating the Pareto surface in nonlinear multicriteria optimization problems. SIAM Journal on Optimization. 1998;**8**:631-657. DOI: 10.1137/S1052623496307510

Chapter 3

The Modeling, Design, Fabrication, and Application of Biosensor Based on Electric Cell-Substrate Impedance Sensing (ECIS) Technique in Environmental Monitoring

Xudong Zhang, William Wang and Sunghoon Jang

Abstract

In this research, the modeling, design, fabrication, and application of ECIS sensors in environmental monitoring are studied. The ECIS sensors are able to qualify the water toxicity through measuring the cell impedance. A novel mathematical model is proposed to analyze the distribution of electric potential and current of ECIS. This mathematical model is validated by experimental data and can be used to optimize the dimension of ECIS electrodes in order to satisfy environmental monitors. The detection sensitivity of ECIS sensors is analyzed by the mathematical model and experimental data. The simulated and experimental results show that ECIS sensors with smaller radius of working electrodes yield higher impedance values, which improves signal-to-noise ratio, which is more suitable in measuring the cell morphology change influenced by environments. Several ECIS sensors are used to detect the toxicant including, phenol, ammonia, nicotine, and aldicarb, and the decreasing cell impedance indicates the toxic effect. The gradient of measured impedance qualitatively indicates the concentration of toxicants in water.

Keywords: ECIS, biosensor, sensitivity, model, electrodes, design, fabrication

1. Introduction

A large number of the world's population live in areas with high risks of environment. Industrialization and usage of nonbiodegradable and non-eco-friendly material are harmful to the environment. Rapid urbanization, increasing population, and extensive agriculture all are threats to the Earth's supply of fresh water. Clean and reliable drinking water can be guaranteed by periodic and extensive testing. Effectively and efficiently environmental monitoring approaches are necessary. ECIS sensing is one of the techniques among them. The ECIS is becoming an increasingly popular technique, which is able to analyze cell behaviors by measuring the impedance profile spectroscopy [1, 2]. The measured cell impedance provides information about cell morphology and electric properties, including intercellular junction conditions, numbers and densities, attachment, migration, proliferation, invasion, barrier function, membrane capacitance, and cytoplasm conductivity [1–6]. A common ECIS sensor is composed of a working electrode and a counter electrode. Some types of ECIS sensors have a third electrode, the reference electrode, which is used to provide the reference voltage for electrochemical measurements. The traditional ECIS sensors are fabricated on rigid substrate that limits the application in some of dynamically moving environments. Zhang et al. [7, 8] have fabricated the ECIS sensors on stretchable polymer. Such sensors are able to simulate in vitro the dynamic environment of organisms, such as pulsation, bending, and stretching, which enables investigations on cell behavior that undergoes mechanical stimuli in biological tissue [9–12].

The cells, attaching and spreading on the ECIS sensors, behave like an insulating medium after seeding. The insulating medium restricts the ion movement between the electrodes [13, 14]. As a result, the measured impedance increases gradually as more cells attach onto the surface. When the cells form a monolayer on the electrodes, the impedance becomes stabilized. The impedance may fluctuate slightly due to cell attachment migration, deformation, and detachment [9, 15–18]. Some chemical, biological, or physical stimuli on measured cells will influence the impedance response due to the changes in cell monolayer caused by cell-cell interactions, cell-substrate interactions, or changing cell electrical properties [2, 9]. Recently, the application of ECIS sensors has been extended to cell-based assays and toxicity study [18].

The ECIS sensors have different configurations including working electrode dimensions, counter electrode dimensions, and distance between electrodes. However, the relationship between the electrode configuration and detection sensitivity has not been further studied. A study on detection sensitivity of ECIS sensors is meaningful for sensor design, fabrication, and applications.

Detection sensitivity is critical in the applications of ECIS sensors, which depends on sensor configuration, such as electrode dimension and the distance between the electrodes [19]. Wang et al. studied the detection sensitivity of ECIS sensors only with interdigital electrodes [20]. Several mathematical models have been introduced to analyze the relationship between measured cell impedance and cell morphology and behaviors [1, 2, 10, 21–28]. In those models, cell membrane and cell cytoplasm were assumed to be capacitors and resistors, respectively, and cell impedance was calculated as a combination of the capacitors and resistors [24–28]. However, the current may switch from one path to another or creating a hybrid path in reality, which was considered by some models [1, 2, 10, 14]. Nevertheless, these models assumed that the current flows radially between the substratum and the ventral surface of the cell, and the electric potential is constant inside the cell. However, the electric potential cannot be assumed to be constant inside the cell if the current flows through the entirety of the cell. This assumption is invalidated by Ohm's law.

In this study, the influence of ECIS sensor configuration on detection sensitivity and the analysis of paths of current flow of ECIS have been carried out for improving the detection sensitivity, design, and application of ECIS sensors. The ECIS sensors are optimized for water toxicity testing. Several ECIS sensors are used to perform the toxicity testing in detecting the toxic effects from phenol, ammonia, nicotine, and aldicarb, and the impedance response successfully indicate the toxic effect. The gradient of measured impedance qualitatively is related to the concentration of toxicants.

2. The mathematical model of electric cell-substrate impedance sensing (ECIS)

In order to monitor the environments effectively, systematically analyzing the relationship between the electric properties of measured subjects and output of ECIS sensors are needed. In this section, a model related to electric field distribution of ECIS sensing, which can be used in quantifying the ECIS sensor measurements, is created with a partial differential equation. The model of ECIS is established in cylindrical coordinates (r, θ, z) as shown in **Figure 1** and simplified into polar coordinates (r, z) due to its axisymmetric property.

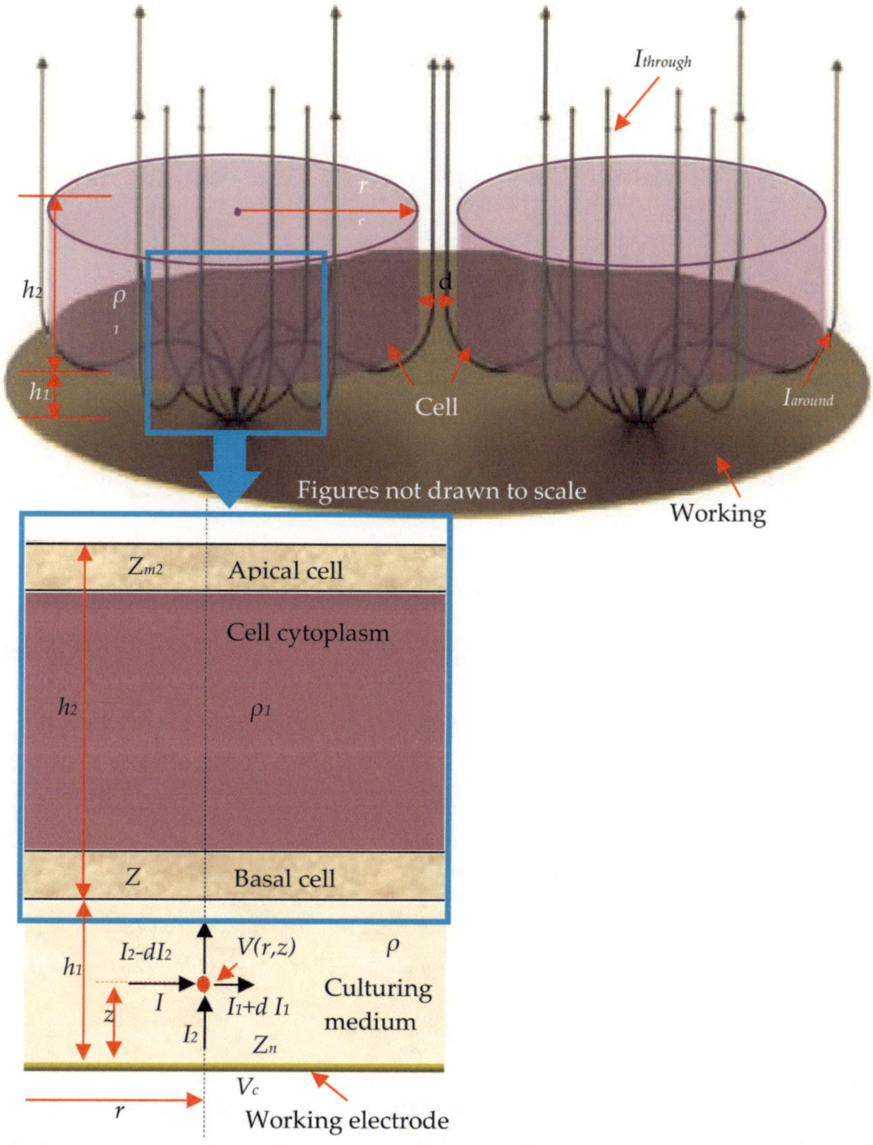

Figure 1.
Illustration of cell impedance sensing on a working electrode. The electric potential at the coordinate (r, z) is V (r, z). ρ and ρ1 are the resistivity of the cell culture medium and cytoplasm respectively. Zm1, Zm2 and Zn are the specific impedance of the basal, apical cell membrane, and electrode-electrolyte interface respectively (in Ωm). h1 is the average distance between the ventral surface of cell and electrode-electrolyte interface. h2 is the average thickness of the cell. d is the average horizontal distance between adhesive cells. Vc is the electrical potential on the working electrode.

The governing equation of electric field distribution of ECIS sensing (as shown in Eq. (1)) can be obtained from the differential form of Ohm's law between electric potential and current (as shown in Eq. (2)), Kirchhoff's circuit law at a point of interest (r, z) (as shown in Eq. (3)), and the gradient of electric potential (as shown in Eq. (4)). The solution of the governing equation is shown in Eq. (5), which is the same as the solution in Giaever et al. ECIS model when the variable z is held as constant [1, 2, 23, 29]. The detailed information about the mode can be referred to in [19]. These three coefficients A, D, and c are calculated as A = −2.3, D = 3.3, and c = 4749.83 by using the parameters listed in [19, 30–36].

$$\frac{2\pi z}{-\rho}\left(\frac{\partial V}{\partial r}+r\frac{\partial^2 V}{\partial r^2}\right)-\frac{2\pi r}{-\rho}\left(\frac{\partial V}{\partial z}\right)=0 \tag{1}$$

$$\rho\left(\frac{I_1}{2\pi rz}e_r+\frac{I_2}{\pi r^2}e_z\right)=E \tag{2}$$

$$I_1+I_2=I_1+dI_1+I_2-dI_2 \tag{3}$$

$$\frac{\partial V}{\partial r}e_r+\frac{\partial V}{\partial z}e_z=-E \tag{4}$$

$$V(r,z)=AI_0(2cr)e^{2c^2z^2}+D \tag{5}$$

where ρ is the resistivity of the cell culture medium (electrolyte); I_1 and I_2 are the current flowing through the point (r,z) in r and z directions, respectively; e_r and e_z are the unit vectors of the r and z directions; E is the electric field at any point (r,z); V is the electric potential at the point (r, z); and dI_1 and dI_2 are the infinitesimally small currents of I_1 and I_2. dI_1 and dI_2 have the same sign; $I_0(2cr)$ is the modified Bessel function of the first kind; A, D, and c are the coefficients of solution $V(r,z)$.

2.1 The calculated impedance of a single cell

In this model, the impedance of a single cell ($Z_{single\ cell}$) is able to be calculated by dividing the electric potential difference between the apical $V(r_c, h_1)$ and ventral surfaces of a single cell $V(r_c, h_1 + h_2)0$ by the total current flowing through and around the cell, as shown in Eq. (6).

$$Z_{single\ cell}=\frac{V(r_c,h_1)-V(r_c,h_1+h_2)}{I_2+I_j}$$
$$=\frac{\left(\rho_1 h_2\sqrt{\sigma^2+(2\pi f\varepsilon\varepsilon_0)^2}+2t\right)(2I_0+2cr_cI_1)}{2\pi\left[r_c{}^2I_0\sqrt{\sigma^2+(2\pi f\varepsilon\varepsilon_0)^2}+\frac{r_cd}{\rho\alpha h_2}\left(\rho_1 h_2\sqrt{\sigma^2+(2\pi f\varepsilon\varepsilon_0)^2}+2t\right)(2I_0+2cr_cI_1)\right]} \tag{6}$$

where I_2 is the current flowing through a single cell, I_j is the current flowing through the intercellular junction gap; h_1 is the average distance between the ventral surface of cell and electrode-electrolyte interface; h_2 is the average thickness of the cell layer; r_c is the average radius of a single cell; f is the measurement frequency; ρ_1 is the resistivity of cell cytoplasm; ε is the relative permittivity of the cell membrane; ε_0 is the vacuum permittivity, which is 8.85×10^{-12} F/m; and t and σ are the thickness and conductivity of the cell membrane, respectively.

2.2 The calculated impedance of a cell monolayer

The impedance of a cell monolayer (Z) is calculated as the sum of the impedance on current path, including the impedance from working electrode $Z_{working}$, counter electrode $Z_{counter}$, and cell culture medium R_s, as shown in Eq. (7).

$$Z = Z_{working} + Z_{counter} + R_s = \left(\frac{1}{S_1} + \frac{1}{S_2}\right)\left[Z_n + \frac{S(Z_{single\ cell} + Z_{cell\_sub})}{n}\right] + R_s \quad (7)$$

where Z_n is the specific impedance of the electrode-medium interface (unit Ωm^2), which can be calculated according to the parameters referred to [19, 37–41]; S_1 and S_2 are the surface areas of the working and counter electrodes, respectively; S is the total surface area of the ECIS sensor, which contains the working electrode, counter electrode, and nonelectrode area; n is the number of cells seeded on the ECIS sensor; R_s is the impedance of the culture medium, which can be calculated according to the parameters referred to [19, 42–47]; $Z_{cell\text{-}sub}$ is the impedance of the culture medium between the electrode-electrolyte interface and ventral surface of cell, which can be calculated by dividing the electric potential difference between the edge and center of a single cell by the total current flowing through and around the cell [19].

3. The design of ECIS sensors for environmental monitoring

The design of ECIS sensors includes the dimensions of working electrodes and counter electrodes, and the distance between them is critical in environmental monitoring because those designing parameters will influence the detection sensitivity of ECIS sensors.

3.1 The design guideline of electrode dimensions of ECIS sensors

The radius of working electrode (R_i) and the distance between the edges of the sensing electrodes (d_{io}) can be optimized by using the mathematical model with the parameters related to cell morphology and electric properties and surrounding culture medium.

3.1.1 The relationship between the radius of working electrode (R_i) and cell impedance

During impedance measurements, ions move through the cell monolayer between the working and counter electrodes which follow many paths. The counter electrodes must have adequate sensing area in order to provide adequate circuit connection. The larger R_i working electrode provides more current paths, which decreases the corresponding impedance. Higher impedance values can improve the data quality of the measured impedance by increasing signal-to-noise ratio, which is useful particularly for sensing small changes in cell behavior. However, the working electrode should not be too small in order to measure adequate number of cells and to guarantee sufficient cell-to-cell contact area.

In this study, the ECIS sensors with R_i from 100 to 400 μm were fabricated to analyze the relationship between R_i and measured cell impedance. **Figure 2** illustrates the cell morphology on those sensors. The simulated cell impedance by using R_i within the same range was also obtained from the mathematical model. The experimental and simulated impedance of cell were shown in **Figure 3**. The

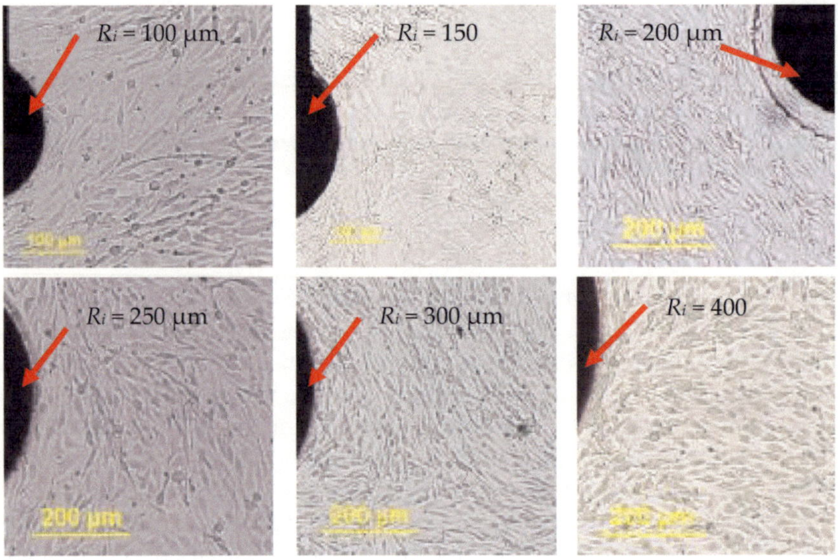

Figure 2.
BAEC monolayer on ECIS sensors with different R_i (d_{io} = 3.5 mm).

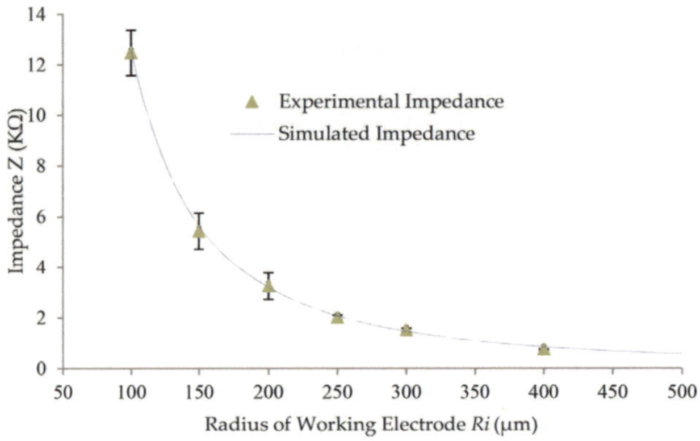

Figure 3.
Relationships between R_i and experimental impedance, and between R_i and simulated impedance at 8000 Hz (n = 4 ~ 6, d_{io} = 3.5 mm).

simulated impedance curve matches the experimental data closely with maximum difference 13.29%, which is acceptable when considering the fluctuation of measured impedance. The consistency of the simulated impedance with the experimental impedance validates this model's ability to optimize the R_i according to the range of measured cell number and expected output impedance level during sensor designing.

3.1.2 The relationship between the distance between the edges of the sensing electrodes (d_{io}) and cell impedance

The distance between the edges of the sensing electrodes (d_{io}) is another factor that should be considered in designing ECIS sensors. **Figure 4** shows the experimental impedance and the simulated impedance with different d_{io}. The average experimental impedance slightly changed from 12.50 to 12.52 KΩ, when d_{io} changed from 1000 to 3500 μm. The simulated impedance was calculated by using Eq. (7).

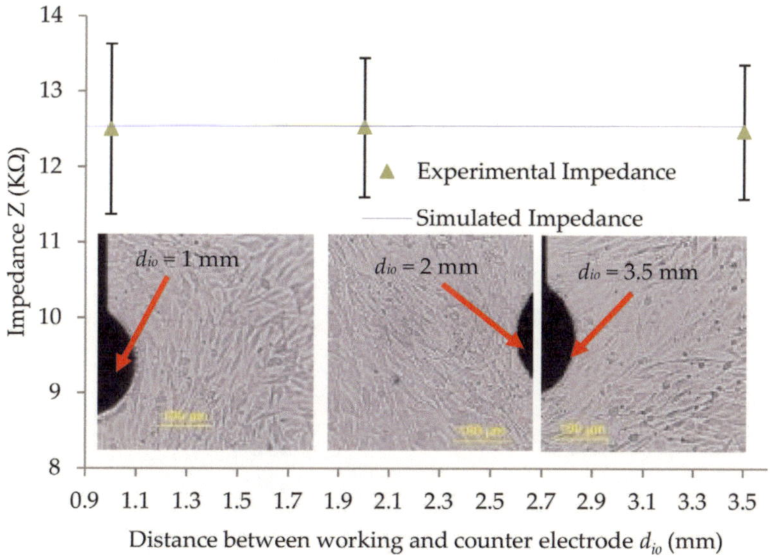

Figure 4.
Relationships between d_{io} and experimental impedance, and between d_{io} and simulated impedance, measured at 8000 Hz ($n = 6 \sim 7$, $R_i = 100\ \mu m$). The three images show the cell morphology of BAECs on the ECIS sensors with d_{io} of 1 mm, 2 mm, and 3.5 mm, respectively.

d_{io} only slightly influences the simulated impedance because the natural logarithm of the quotient of $(R_i + d_{io})$ and R_i makes the influence of d_{io} on simulated impedance more slightly in Eq. (7). The simulated impedance is consistent with the experimental data with maximum difference 0.63%, which validates the model. The experimental and simulated impedance indicates that d_{io} in the range of 1000–3500 μm has only a little influence on the impedance because d_{io} influenced the impedance of medium, which is only a small portion of measured impedance. Thus, d_{io} cannot dramatically influence the measured impedance. However, d_{io} should be large enough to avoid the current bypassing the cell monolayer between sensing electrodes.

3.2 The influence of electrode dimensions on the detection sensitivity of ECIS

Detection sensitivity reflects the fineness of impedance response to the changes of cell behavior in cell-based assay environmental monitoring. The detection sensitivity of ECIS sensors is influenced by R_i. According to the previous experimental results, ECIS sensors with R_i larger than 200 μm do not respond sensitively and quickly on cell morphology changes. So, ECIS sensors with R_i of 100 and 150 μm were fabricated to study the influence of R_i on the detection sensitivity. Cell densities, 90,000, 100,000, and 110,000 cells/cm^2, were used to study the relationship between cell density and impedance. **Figure 5** shows the impedance shifts versus the cell density changes with R_i of 100 and 150 μm. **Figure 6** shows the corresponding cell morphology on different ECIS sensors. When the cell density changes are 10,000 cells/cm^2 (from 90,000 to 100,000 cells/cm^2), the impedance increased 597 and 350 Ω for the sensors with R_i of 100 and 150 μm, respectively. When the cell density changes are 20,000 cells/cm^2 (from 90,000 to 110,000 cells/cm^2), the impedance increased 1336 and 880 Ω for the sensors with R_i of 100 and 150 μm, respectively. The experimental results indicate that the sensors with larger R_i illustrate less impedance changes with the same amount of cell density changes. Therefore, the sensors with smaller R_i are able to detect more

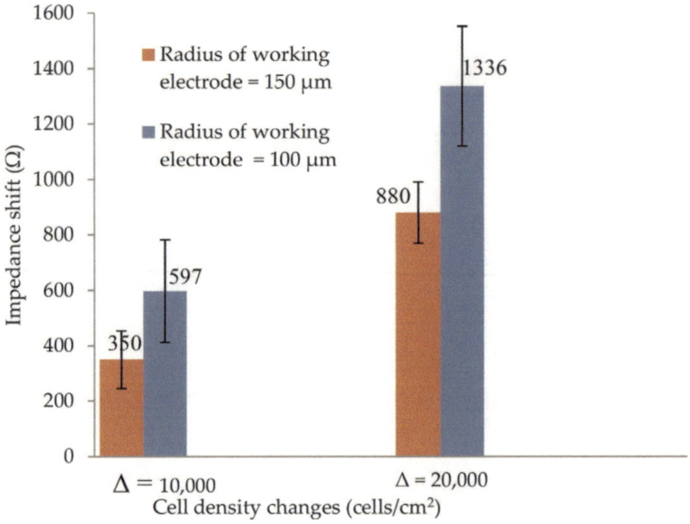

Figure 5.
Impedance shifts to cell density changes with sensors' R_i of 100 μm and 150 μm (n = 3). The cell density change from 90,000 cells/cm^2 to 100,000 or 110,000 cells/cm^2.

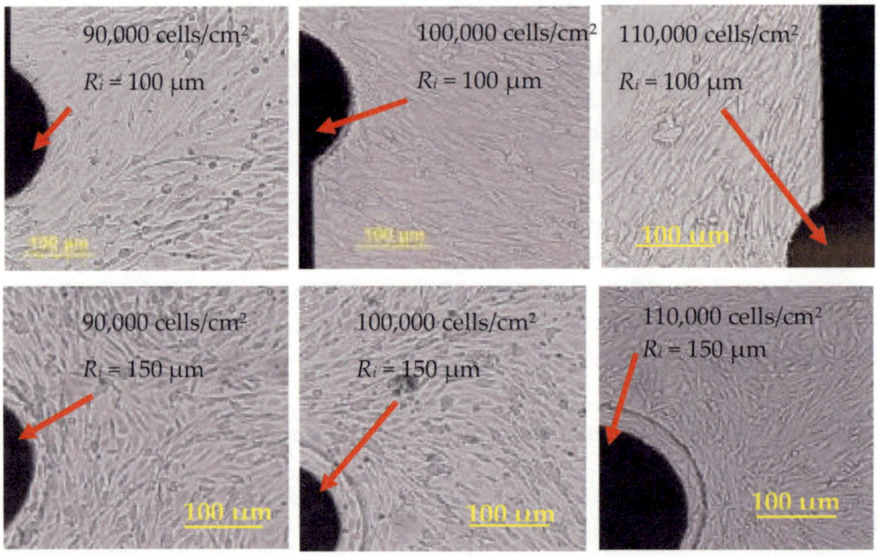

Figure 6.
Cell morphology with 90,000, 100,000, and 110,000 cells/cm^2 cell densities on ECIS sensors (R_i = 100 μm and R_i = 150 μm).

sensitive changes in cell density. Therefore, ECIS sensors with smaller dimension working electrodes illustrate better detection sensitivity on changes in cell density. Another benefit is that smaller R_i requires fewer cells in cell-based assays.

Based on the analysis above, the ECIS sensors with R_i of 100–125 μm and d_{io} of 3.5 mm are preferred in environmental monitoring because R_i of 100–125 μm will allow the ECIS sensors to be sensitive to sense the cell morphology changes due to environment influence and own good anti-interference ability. The area of counter electrodes should be as large as possible to guarantee sufficient contact area between electrode and cells. d_{io} of 3.5 mm is enough to avoid the current bypassing the cell layer in ECIS measurements.

4. Fabrication of ECIS sensor arrays

The fabrication of ECIS sensors can follow different photolithography techniques. The substrates are usually nonconductive materials, includes glass, printed circuit board (PCB) [1–4, 19, 23], and polymer including polydimethylsiloxane (PDMS) [9] and polycarbonate [1–3, 18, 19]. The ECIS arrays were fabricated on glass by thin film deposition and lift-off photolithography technique, as shown in **Figure 7**. Initially, the photoresist AZ5214E (MicroChemicals, Somerville, NJ) was coated on glass slides with spinning coater at 2000 rpm. After baking on hotplate at 110°C for 50 seconds, the coated photoresist was exposure to ultraviolet (UV) light. Then, a reversal bake is carried out at 120°C for 2 minutes. Finally, UV light with intensity larger than 200 mJ/cm^2 was exposure on the photoresist pattern. The electrode pattern was created after immersing the slides with photoresist in the AZ 100 Remover (MicroChemicals, Somerville, NJ). The remover is able to dissolve the photoresist without the first exposure (image reverse). A 20-nm-thick chromium (Cr) followed by a 150-nm-thick gold (Au) was coated on the substrate to form the sensor's electrodes by thermal evaporation. The sensing electrodes were formed after the lift-off process. Then, the photoresist SU-8 (MicroChem, Westborough, MA) was used to cover the substrate except the sensing areas. The sensor arrays were treated with 95% sulfuric acid at 60°C for 15 seconds [48] followed by washing with deionized water (DI) and then treated with 8% (3-aminopropyl) triethoxysilane (APTES) at 50°C for 2 hours to improve the surface biofunctionality. Finally, cell culture wells (Lab-Tek 8-well culture wares) were glued onto the sensor array. **Figure 8** shows the fabricated ECIS sensor array and its configuration. R_i is the radius of the working electrode, R_{co} is the outer radius of the

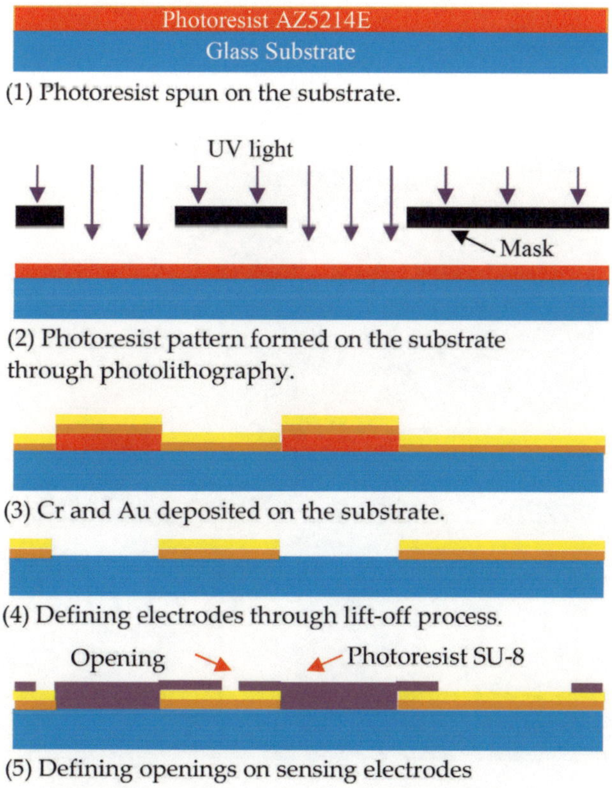

Figure 7.
Illustration of ECIS sensor fabrication.

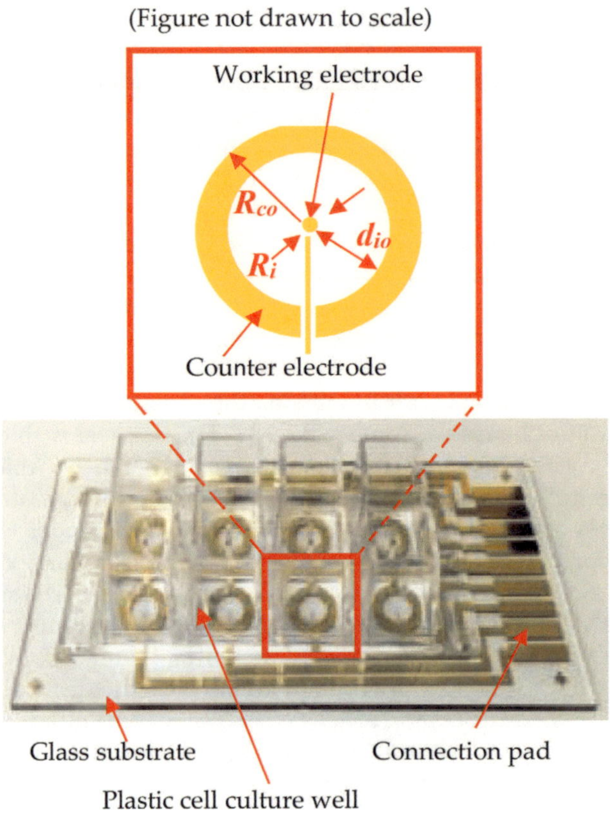

Figure 8.
The array of eight ECIS sensors.

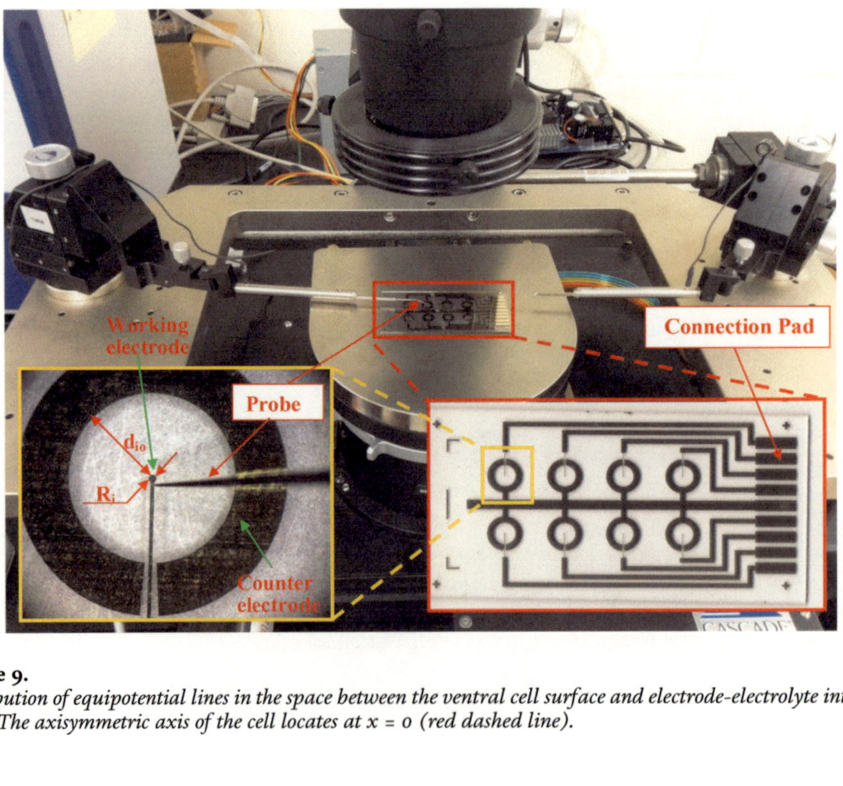

Figure 9.
Distribution of equipotential lines in the space between the ventral cell surface and electrode-electrolyte interface layer. The axisymmetric axis of the cell locates at x = 0 (red dashed line).

counter electrode, d_{io} is the distance between the edges of the electrodes, and S_1 and S_2 are the areas of the working and counter electrodes, respectively.

The inherent impedance of the Au/Cr electrodes of fabricated ECIS sensors is measured by microwave probe station (Cascade Microtech Inc., Beaverton, OR) and impedance analyzer (Agilent 4294) as shown in **Figure 9**. The maximum inherent impedance was 19 Ω at 8000 Hz, which is much lower than the measured cellular impedance of thousands of ohms. Thus, the inherent impedance of the sensor can be neglected.

5. The application of ECIS sensors in environmental monitoring

5.1 Cell culture and preparation

Bovine aortic endothelial cells (BAECs, VEC Technologies, Rensselaer, NY) were used in this study. The BAECs were cultured in minimum essential medium (MEM, GIBCO, Grand Island, NY) with 10% fetal bovine serum (FBS, GIBCO, Grand Island NY) under standard mammalian cell culturing conditions (37°C and 5% CO_2). Confluent BAEC were trypsinized to detach the cells from the cell culture flasks to prepare the cell suspension. Then, the cell suspension was centrifuged on the bottom of centrifuge tube followed by aspirating off the upper supernatant. Finally, certain amount of cell culture medium was added into centrifuge tube to prepare specific concentration of the cell suspension.

5.2 Toxicant preparation

This study investigated the toxicant detection by using the ECIS sensors. The toxicants used in this study are phenol (RICCA, Arlington, TX), ammonia (Acros Organics, Fair Lawn, NJ), nicotine (Fisher Scientific, Hanover Park, IL), and aldicarb (SPEX CertiPrep, Metuchen, NJ). All the toxicants were diluted with Dulbecco's phosphate-buffered saline (DPBS, Mediatech, Inc., Manassas, VA). The osmolarity of diluted toxicant solution was considered to be in the suitable range for cell culture because the small volume of toxicants added into DPBS will not change the concentration of essential ingredients of DPBS dramatically.

5.3 Experimental system setup

Impedance analyzer Agilent 4294 and ECIS measurement system (Applied Biophysics, Troy, NY) was used to measure the cell impedance. The AC signal applied to the cells was monitored by using Tektronix oscilloscope DPO2014B. Two MAXIM DG408 Multiplexers, controlled by an NI USB-6008 multifunction data acquisition card, were used as a 16-channel multiplexer between the impedance analyzer and the sensor arrays. The NI USB-6008 and Agilent 4294 were controlled by LabVIEW programs to perform the data acquisition shown as **Figure 10**. The ECIS sensor arrays, covered with BAECs on the sensing electrodes, were kept in an incubator with 37°C and 5% CO_2 during the impedance measurement.

5.4 Optimization of cell seeding density and measurement frequency

The cell seeding density and measurement frequency are need to be optimized to obtain reasonable measurement results. BAECs were seeded with different cell densities of 150,000, 125,000, and 100,000 cells/cm^2 on ECIS sensor. The impedance values were recorded and normalized in the initial 46 hours after seeding onto the ECIS sensor array, as shown in **Figure 11**. The morphology of cells with seeding density 125,000 cells/cm^2 at different time points was also shown in **Figure 11**. The cells gradually spread on the surface of ECIS sensors after seeding and eventually form a monolayer with stable impedance. The cell impedance gradually increased in the initial 8–20 hours, which indicates the initial formation of a loose monolayer and stable up to the end of the impedance measurements. The cell morphology was

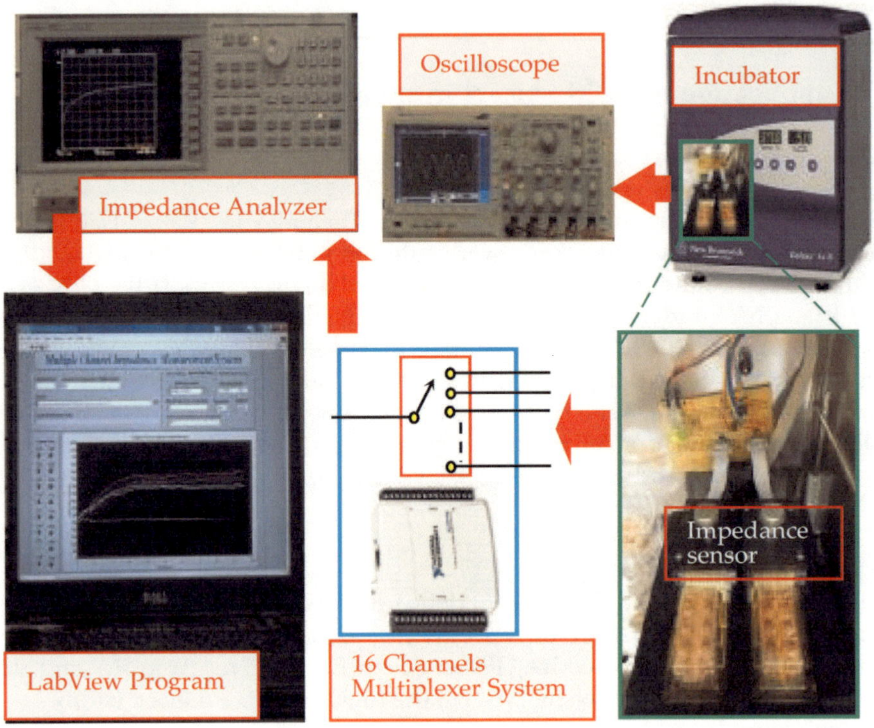

Figure 10.
Experimental setup of cell impedance measurement.

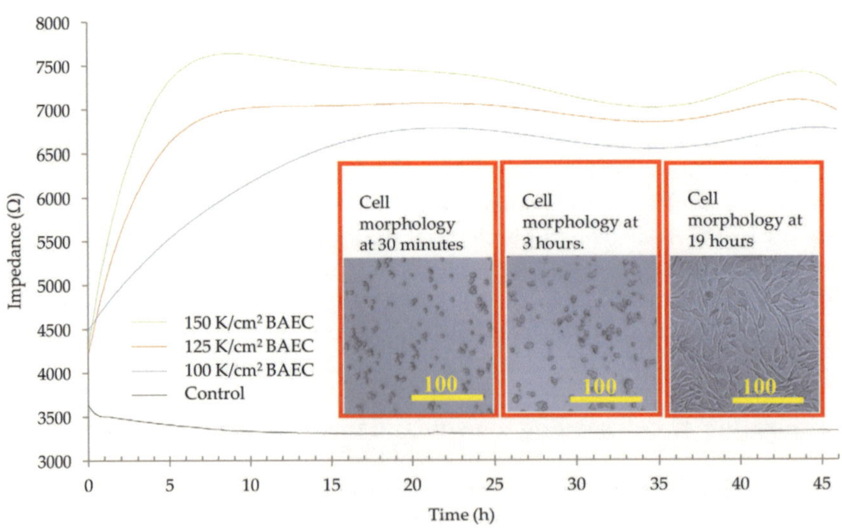

Figure 11.
Impedance response of BAECs measured by an ECIS sensor array at 8000 Hz and the image of cell morphology.

checked under microscope frequently. The corresponding impedance readings were used to represent the impedance of the cell monolayer for cell-based assays. In **Figure 11**, the impedance of cell monolayer with higher seeding densities increases more rapidly than cells with lower seeding densities because higher seeding densities allow the cells to have tighter and stronger intercellular junctions and the corresponding ion insulating abilities are better. The impedance of cells with the highest seeding density, 150,000 cells/cm^2, decreased after initial formation (around 8 hours) of cell monolayer due to the cell movement on the surface of ECIS sensors. Also, the impedance of cells with 150,000 cells/cm^2 seeding density is not stable as those with 125,000 and 100,000 cells/cm^2 seeding densities. The cells with a seeding density of 100,000 cells/cm^2 need 20 hours to be confluent and have low impedance compared with those with higher seeding densities. Hence, the cell seeding density, 125,000 cells/cm^2, was chosen as the preferred seeding density in the toxicity testing.

The optimal measurement frequency allows the sensors to obtain the largest difference in measured impedance between a sample with and without cells [19]. In this study, the impedance of cell monolayer was measured with different frequencies from 500 Hz to 64 kHz. The optimal measurement frequency was optimized to be 8000 Hz in experimental measurements.

5.5 Toxicity testing

The ECIS sensors need to be prepared before the toxicity testing. ECIS sensors were cleaned by oxygen plasma to provide a sterilized surface for cell seeding. Then phosphate-buffered saline (PBS, GIBCO, Grand Island, NY) was used to clean the sensor surface again. Before cell seeding, 30 μg/ml fibronectin (GIBCO, Grand Island, NY) was coated on the surface of the sensors to improve cell attachment. BAECs were seeded onto each sensor with a seeding density of 125,000 cells/cm^2. The cell morphology was checked under microscope. The selected toxicants were introduced to each well to perform toxicity testing after monolayer formation. **Figure 12** shows normalized impedance response after introducing 0.1 and 0.2 mM aldicarb and the cell morphology after introducing 0.2 mM aldicarb. Some of the cells detached from the substrate. The normalized impedance decreases to 0.84 and

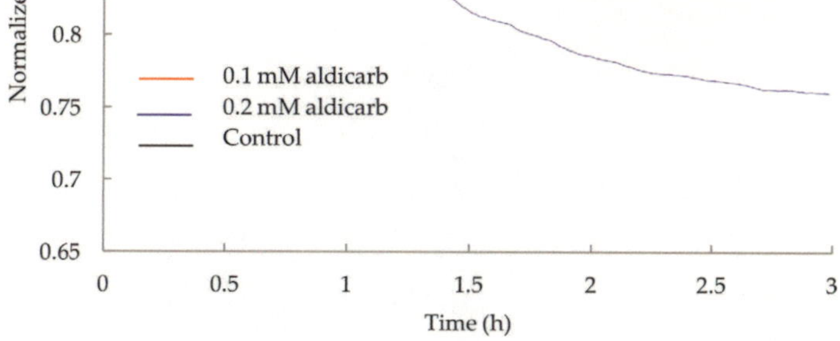

Figure 12.
The normalized impedance of BAEC exposed to aldicarb.

0.76 times its original impedance value within 3 hours when treated with 0.1 and 0.2 mM aldicarb, respectively. The cell morphology changed and even detached from the sensors. **Figure 13** shows the normalized impedance response after introducing 0.1 and 0.2 mM phenol as toxicant. The BAEC detached from substrate after introducing 0.2 mM phenol. The decreasing impedance curves indicate the toxic effect on BAECs. The normalized impedance values rapidly decreased to 0.80 and 0.74 times its original impedance value within 2 hours when treated with 0.1 and 0.2 mM phenol, respectively. The image shows the cells obviously detached from the sensor. **Figure 14** shows the normalized impedance response after introducing 2 and 5 mM ammonia as toxicant. Those lines shows that the normalized impedance values rapidly decreased to 0.78 and 0.68 times its original impedance value within 1 hour when treated with 2 and 5 mM ammonia, respectively. The image shows the cell morphology after 1 hour after introducing ammonia. The cells morphology changed and very easily detached from the sensor substrate. **Figure 15** shows the

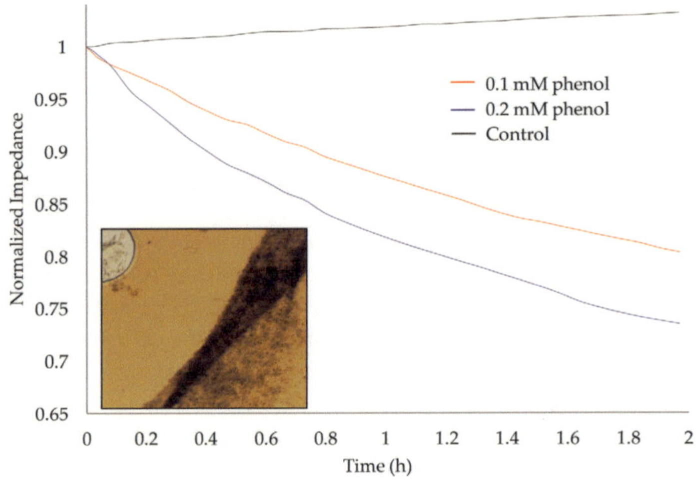

Figure 13.
The normalized impedance of BAEC exposed to phenol.

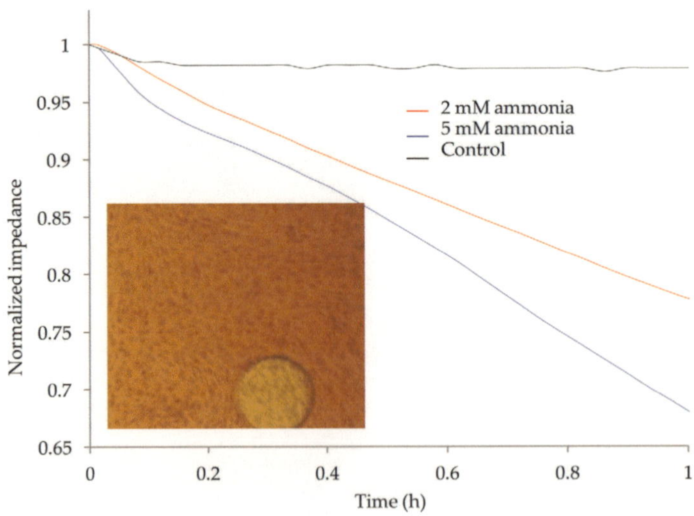

Figure 14.
The normalized impedance of BAEC exposed to 2 mM and 5 mM ammonia, and the image shows the BAEC morphology after exposure to 5 mM of ammonia.

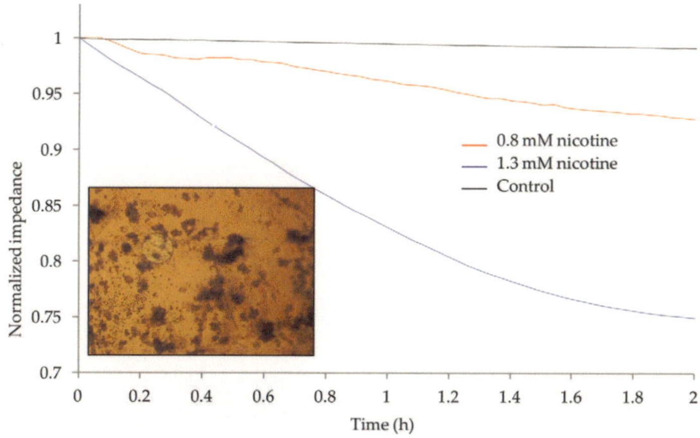

Figure 15.
The normalized impedance of BAEC exposed to 0.8 mM and 1.3 mM nicotine. The image shows the cell morphology after exposure to 1.3 mM nicotine.

normalized impedance of BAEC after exposing to 0.8 and 1.3 mM nicotine as toxicant. The normalized impedance rapidly decreased to 0.92 and 0.75 times its original value within 2 hours when treated with 1.3 and 0.8 mM nicotine, respectively. The image shows the BAEC morphology after exposed to 1.3 mM nicotine. Most of the cell detached from the sensor due to the toxic effect of nicotine.

The cell morphology and decreasing impedance curves indicate the toxic effect and the effectiveness of ECIS sensing on environmental monitoring within short period of time. Different concentrations of toxicants are qualified according to the gradients of normalized impedance. ECIS sensing technique is able to perform environmental monitoring effectively and efficiently compared with other approaches.

6. Conclusions

In this study, the biosensors based on ECIS sensing technique were used to monitor and measure the environmental toxicants, including the phenol, ammonia, nicotine, and aldicarb. A model, validated by experimental results, was created to analyze the electric potential distribution of ECIS sensing and guide the designing, especially the sensing area of sensor electrodes. The detection sensitivity of ECIS sensors was optimized. The experimental results show that ECIS sensors are capable to detect and qualify the environmental toxicants rapidly. The concentration of toxicants can be indicated from the gradients of normalized cell impedance.

Acknowledgements

We appreciate Dr. Ioana Voiculescu's and Andres Rivera's support in this study.

Author details

Xudong Zhang[1*], William Wang[2] and Sunghoon Jang[3]

1 City University of New York, New York, USA

2 Stanford University, Stanford, USA

3 New York City College of Technology of City University of New York, USA

*Address all correspondence to: xzhang19@citymail.cuny.edu

References

[1] Giaever I, Keese C. Monitoring fibroblast behavior in tissue culture with an applied electric field. Proceedings of the National Academy of Sciences. 1984; **81**:3761-3764

[2] Giaever I, Keese CR. Use of electric fields to monitor the dynamical aspect of cell behavior in tissue culture. IEEE Transactions on Biomedical Engineering. 1986;**33**:242-247

[3] Mitra P, Keese CR, Giaever I. Electric measurements can be used to monitor the attachment and spreading of cells in tissue culture. BioTechniques. 1991;**11**: 504-510

[4] Tiruppathi C, Malik AB, Del Vecchio PJ, Keese CR, Giaever I. Electrical method for detection of endothelial cell shape change in real time: Assessment of endothelial barrier function. Proceedings of the National Academy of Sciences. 1992;**89**:7919-7923

[5] Pei Z, Keese CR, Giaever I, Kurzawa H, Wilson DE. Effect of the pSV2-neo plasmid on NIH 3T3 cell motion detected electrically. Experimental Cell Research. 1994;**212**:225-229

[6] Keese CR, Giaever I. A biosensor that monitors cell morphology with electrical fields. IEEE Engineering in Medicine and Biology Magazine. 1994; **13**:402-408

[7] Voiculescu I, Zhang X, Cell-on-chip stretchable platform for mammalian cells with integrated impedance spectroscopy technique. In: Google Patents; 2017

[8] Zhang X, Nordin AN, Li F, Voiculescu I. Stretchable lab-on-chip device with impedance spectroscopy capability for mammalian cell studies. In: Proceedings of the ASME 2016 International Mechanical Engineering Congress and Exposition; American Society of Mechanical Engineers; 2016. pp V010T13A005-V010T13A005

[9] Zhang X, Wang W, Li F, Voiculescu I. Stretchable impedance sensor for mammalian cell proliferation measurements. Lab on a Chip. 2017;**17**: 2054-2066

[10] Zhang X. Stretchable impedance spectroscopy sensor for mammalian cell studies [thesis]. The City College of New York; 2016

[11] Zhang X, Li F, Lee K, Voiculescu I. Lab-on-chip stretchable impedance spectroscopy device for mammalian cells studies. In: Proceedings of 19th International Conference on the Solid-State Sensors, Actuators and Microsystems (TRANSDUCERS); IEEE; 2017. pp. 1563-1566

[12] Zhang X, Petrissans R, Li F, Voiculescu I. Stretchable impedance spectroscopy sensor for mammalian cells impedance measurements. In: Proceedings of the ASME 2014 International Mechanical Engineering Congress and Exposition; American Society of Mechanical Engineers; 2014. pp. V010T13A005-V010T13A005

[13] Voiculescu I, Li F, Liu F, Zhang X, Cancel LM, Tarbell JM, et al. Study of long-term viability of endothelial cells for lab-on-a-chip devices. Sensors and Actuators B: Chemical. 2013;**182**: 696-705

[14] Ayliffe HE, Frazier AB, Rabbitt RD. Electric impedance spectroscopy using microchannels with integrated metal electrodes. Journal of Microelectromechanical Systems. 1999; **8**:50-57

[15] Han A, Moss E, Frazier AB. Whole cell electrical impedance spectroscopy for studying ion channel activity. In: Proceedings of the 13th International

Conference on Solid-State Sensors, Actuators and Microsystems, 2005. Digest of Technical Papers. TRANSDUCERS'05; IEEE; 2005. pp. 1704-1707

[16] Han A, Yang L, Frazier AB. Quantification of the heterogeneity in breast cancer cell lines using whole-cell impedance spectroscopy. Clinical Cancer Research. 2007;**13**:139-143

[17] Solly K, Wang X, Xu X, Strulovici B, Zheng W. Application of real-time cell electronic sensing (RT-CES) technology to cell-based assays. Assay and Drug Development Technologies. 2004;**2**: 363-372

[18] Zhang X, Li F, Nordin AN, Tarbell J, Voiculescu I. Toxicity studies using mammalian cells and impedance spectroscopy method. Sensing and Bio-Sensing Research. 2015;**3**:112-121

[19] Zhang X, Wang W, Nordin AN, Li F, Jang S, Voiculescu I. The influence of the electrode dimension on the detection sensitivity of electric cell-substrate impedance sensing (ECIS) and its mathematical modeling. Sensors and Actuators B: Chemical. 2017;**247**: 780-790

[20] Wang L, Wang H, Mitchelson K, Yu Z, Cheng J. Analysis of the sensitivity and frequency characteristics of coplanar electrical cell-substrate impedance sensors. Biosensors and Bioelectronics. 2008;**24**:14-21

[21] Öz S, Maercker C, Breiling A. Embryonic carcinoma cells show specific dielectric resistance profiles during induced differentiation. PLoS One. 2013;**8**:e59895

[22] Gentet LJ, Stuart GJ, Clements JD. Direct measurement of specific membrane capacitance in neurons. Biophysical Journal. 2000;**79**:314-320

[23] Lo C-M, Keese CR, Giaever I. Impedance analysis of MDCK cells measured by electric cell-substrate impedance sensing. Biophysical Journal. 1995;**69**:2800

[24] Asphahani F, Zhang M. Cellular impedance biosensors for drug screening and toxin detection. The Analyst. 2007;**132**:835-841

[25] Spira ME, Hai A. Multi-electrode array technologies for neuroscience and cardiology. Nature Nanotechnology. 2013;**8**:83-94

[26] Li H, Zou Q, Zou L, Wang Q, Su K, Hu N, et al. Detection of cardiovascular drugs and marine toxins using a multifunctional cell-based impedance biosensor system. Analytical Methods. 2015;**7**:7715-7723

[27] Wang J, Wu C, Hu N, Zhou J, Du L, Wang P. Microfabricated electrochemical cell-based biosensors for analysis of living cells in vitro. Biosensors. 2012;**2**:127-170

[28] Anderson SE, Bau HH. Electrical detection of cellular penetration during microinjection with carbon nanopipettes. Nanotechnology. 2014;**25**: 245102

[29] Giaever I, Keese CR. Micromotion of mammalian cells measured electrically. Proceedings of the National Academy of Sciences. 1991;**88**: 7896-7900

[30] Chen J, Zheng Y, Tan Q, Shojaei-Baghini E, Zhang YL, Li J, et al. Classification of cell types using a microfluidic device for mechanical and electrical measurement on single cells. Lab on a Chip. 2011;**11**:3174-3181

[31] Chen J, Zheng Y, Tan Q, Zhang YL, Li J, Geddie WR, et al. A microfluidic device for simultaneous electrical and mechanical measurements on single cells. Biomicrofluidics. 2011;**5**:014113

[32] Park S, Zhang Y, Wang T-H, Yang S. Continuous dielectrophoretic

bacterial separation and concentration from physiological media of high conductivity. Lab on a Chip. 2011;**11**: 2893-2900

[33] McNutt NS, Weinstein RS. The ultrastructure of the nexus: A correlated thin-section and freeze-cleave study. The Journal of Cell Biology. 1970;**47**: 666-688

[34] Bohnert J. Effects of Time-Varying Magnetic Fields in the Frequency Range 1 kHz to 100 kHz upon the Human Body: Numerical Studies and Stimulation Experiment. Karlsruhe: KIT Scientific Publishing; 2014

[35] Iwanaga Y. Cell-substrate distance measurement in correlation with distribution of adhesion molecules by fluorescence microscopy [thesis]. Technische Universität München; 2000

[36] Lang Q, Wu Y, Ren Y, Tao Y, Lei L, Jiang H. AC electrothermal circulatory pumping chip for cell culture. ACS Applied Materials & Interfaces. 2015;**7**: 26792-26801

[37] Randles JEB. Kinetics of rapid electrode reactions. Discussions of the Faraday Society. 1947;**1**:11-19

[38] Schwan H. Electrode polarization impedance and measurements in biological materials. Annals of the New York Academy of Sciences. 1968;**148**: 191-209

[39] Warburg E. Ueber das Verhalten sogenannter unpolarisirbarer Elektroden gegen Wechselstrom. Ann. Phys. 1899;**303**:493-499

[40] Lvovich VF. Impedance Spectroscopy: Applications to Electrochemical and Dielectric Phenomena. Hoboken: John Wiley & Sons; 2012

[41] Nguyen TA, Yin T-I, Reyes D, Urban GA. Microfluidic chip with integrated electrical cell-impedance sensing for monitoring single cancer cell migration in three-dimensional matrixes. Analytical Chemistry. 2013;**85**: 11068-11076

[42] Franks W, Schenker I, Schmutz P, Hierlemann A. Impedance characterization and modeling of electrodes for biomedical applications. IEEE Transactions on Biomedical Engineering. 2005;**52**:1295-1302

[43] Henisch HK. Rectifying Semi-Conductor Contacts. Oxford: Clarendon Press; 1957

[44] Kovacs GT. Introduction to the Theory, Design, and Modeling of Thin-Film Microelectrodes for Neural Interfaces. San Diego: Academic Press; 1994. pp. 121-165

[45] Schroder DK. Semiconductor Material and Device Characterization. Hoboken: John Wiley & Sons; 2006

[46] Joye N, Schmid A, Leblebici Y. An electrical model of the cell-electrode interface for high-density microelectrode arrays. In: Proceedings of the 30th Annual International Conference of the IEEE Engineering in Medicine and Biology Society; IEEE; 2008. pp. 559-562

[47] Alberto Yf, Alberto O. Cell Biometrics Based on Bio-Impedance Measurements. In: Chetty G, editors. Advanced biometric technologies. London: InTech; 2011 p. 343-366. DOI: 10.5772/21742

[48] Xue P, Bao J, Chuah YJ, Menon NV, Zhang Y, Kang Y. Protein covalently conjugated SU-8 surface for the enhancement of mesenchymal stem cell adhesion and proliferation. Langmuir. 2014;**30**:3110-3117

Chapter 4

Electrochemical Biosensors Containing Pure Enzymes or Crude Extracts as Enzyme Sources for Pesticides and Phenolic Compounds with Pharmacological Property Detection and Quantification

*Flavio Colmati, Lívia Flório Sgobbi,
Guilhermina Ferreira Teixeira, Ramon Silva Vilela,
Tatiana Duque Martins and Giovanna Oliveira Figueiredo*

Abstract

Biosensors are chemical sensors in which the recognition system is based on a biochemical mechanism. They perform the specific component detection in a sample through an appropriate analytical signal. Enzyme-based biosensors are the most prominent biosensors because of their high specificity and selectivity; besides being an alternative to the common immunosensors, they are more expensive and present a limited binding capacity with the antigen depending on assay conditions. This chapter approaches the use of enzymes modified electrodes in amperometric biosensing application to detect and quantify pesticides and phenolic compounds with pharmacological properties, as they have been a promising analytical tool in environmental monitoring. These biosensors may be prepared from pure enzymes or their crude extracts. Pure enzyme-based biosensors present advantages as higher substrate specificity and selectivity when compared to crude extract enzymatic biosensors; nevertheless, the enzyme high costs are their drawbacks. Enzymatic crude extract biosensors show lower specificity due to the fact that they may contain more than one type of enzyme, but they may be obtained from low-cost fabrication methods. In addition, they can contain enzyme cofactors besides using the enzyme in its natural conformation.

Keywords: polyphenol oxidase, peroxidase, acetylcholinesterase, crude extracts, biosensors, pesticides, phenolic compounds, environmental enzymatic biosensors

1. Introduction

Chemical sensors and biosensors are devices used in detection and quantification of an analyte by converting its concentration into an analytical signal. Advances in sensor technology have been important for the enrollment of sensing methods in several applications. Chemical sensor operates based on chemical principles, where the analytical signal emerges as a result of a chemical reaction between the analyte and a specific sensitive layer. Electrochemical sensors are able to detect H2, consisting of Pt, Pd, Au, Ag, and metal oxides, as reported by Korotcenkov et al. [1]. These capabilities are expected to be performed by biosensors as well, which are sensors that present a biological recognition element integrated with the transducer. The most popular biosensors are the enzymatic-based ones, successfully represented by the glucose biosensors. Biosensors have become an attractive analytical instrument for environmental monitoring because there still severe barriers through an effective, fast, and low-cost monitoring of harmful pollutants. Among the hazardous contaminants, phenolic compounds and pesticides represent potential human health and environmental risks. Regarding this, there are several studies reporting the use of horseradish peroxidase (HRP) for phenolic compounds and hydrogen peroxide detection [2]. Enzyme-based biosensors operate by indirectly detecting analytes, through detection of consumption or production of specific compounds in the biochemical reaction progress [3]. Phenolic pollutants are important due to their extensive use in several industrial products and their resulting negative environmental impacts. Also, enzymatic biosensors are applied to detect pesticides, particularly organophosphorus and carbamates. The operation of these devices, primarily designed to quantify those pesticides, is based on the inhibition of enzyme activity by these toxic compounds. Distinctly, the use of enzymes in biosensors for environmental monitoring brings considerable advantages, such as high selectivity and specificity, enhanced sensitivity, catalytic activity, and fast performance [4]. Nevertheless, they present some drawbacks associated with the high costs of obtainment and manipulation processes (extraction, isolation, and purification), denaturation during immobilization on transducer, and activity loss after a period (short shelf life) [4]. However, when enzymatic biosensors are compared to other sensing devices, such as immunosensors, they show superior characteristics because antibodies are more expensive, they do not present catalytic activity, and their binding ability depends on conditions of the assay, such as temperature and pH. Due to their advantages, the use of enzymatic biosensors to monitor environmental pollutants, as well as their applications in pharmacology and in pesticides monitoring will be discussed in this chapter.

2. Phenolic pollutants

Phenolic compounds are present in daily activities, since they are frequently found in vegetables, materials, waste, and water, not mentioning their relevance to several applications, due to their pharmacological and antioxidant properties [4]. Beyond the natural phenolic compounds, the synthetic ones are used in many daily products, such as fragrances, moisturizers, makeup, drugs, processed foods, and plastics, among others [5]. The manufacture and use of these products result in their accumulation in the environment, mostly in water.

Several phenolic compounds have been appointed as endocrine disrupting chemicals (EDCs), defined as "chemical substances or mixtures that interfere in any aspect of the hormonal action of living organisms" [6]. EDCs comprise many chemicals used in industrial activities, such as natural and synthetic hormones, pharmaceuticals, pesticides, and surfactants. Some examples of phenolic compounds appointed as EDCs and related products are shown in **Table 1**.

Phenolic pollutants are worldwide dispersed; they can be transported at long distances by water flows and show high persistence and penetrability [7]. Exposure of aquatic animal species, including fishes and amphibians, to EDCs has been related to be responsible for the observed feminization of many species, which in contrast, diminishes the population of these species. Studies of the exposure effects of humans to EDCs suggest a relation between the development of chronic diseases, such diabetes mellitus type II, obesity, thyroid dysfunction, poor quality sperm in males, and fertility issues [8]. Although, until now, there is no effective confirmation of the effect of EDC exposure to these metabolic anomalies, monitoring the environmental concentration of such substances had been the actual concern of the scientific community. Due to their low cost, selectivity, sensitivity, and fast response, biosensors have been considered a promising alternative to classic analytical methods, such chromatography and nuclear magnetic resonance.

2.1 Enzymatic biosensors

Due their complex structures, enzymes exhibit high selectivity to substrates, being able to detect one substance in multicomponent matrices. This behavior is exploited in analytical devices that present high reproducibility, sensibility, and selectivity, making use of low time-consuming analysis, low-cost equipment, and few or any sample preparation steps [9]. These advantages combined with electrochemical transducers result in cheaper portable and miniaturized biosensors, when compared to other types of transducers, such as optical and piezoelectric [10], which is a great feature for environmental applications.

The electrochemical enzymatic biosensors operate based on the electron transfer between the enzyme active site and the substrate, which is, then, transduced to generate an analytical signal. The electrochemical signal can be of three distinct types: (i) amperometric, in which the electrical current generated in the electron transfer process is measured [11], (ii) conductimetric, in which the change in the electrical conductivity of the environment is measured [12], and (iii) potentiometric, in which the electrochemical potential in the absence of measurable current is measured [13]. The amperometric biosensors are the most used ones, due to their high sensibility. These biosensors require the enzyme immobilization on the electrode surface. The most frequently used methods for enzyme immobilization are noncovalent adsorption, covalent bonding, entrapment, cross-linking, and affinity, and they are discussed below [14].

2.2 Enzyme immobilization on the electrode surface

The noncovalent adsorption immobilization consists of enzyme adsorption on the electrode surface by physical interactions, such as van der Waals forces,

Product class	EDC examples
Drugs (human and animal uses)	Acetaminophen, tetracyclines, salbutamol, morphine
Antimicrobials (food and cosmetics)	Chlorophenols, parabens, triclosan, propyl gallate, tert-butylhydroquinone
Plastics	Bisphenol A (BPA), bisphenol F (BPF)
Steroids	Estradiol, estrone, estriol
Surfactants	Alkylphenols

Table 1.
Phenolic compounds appointed as EDCs and their related products.

hydrogen bonds, and electrostatic interactions [14]. In contrast, in the covalent bonding immobilization, the enzyme is anchored on the electrode surface by multiple covalent bonds between support functional groups and enzymes. The entrapment immobilization on the electrode is the enzyme inclusion in a framework, such as a polymer network, which can be organic or inorganic polymeric matrices. An additional method for enzyme immobilization that provides high stability is the application of a metal-organic framework (MOF) [15]; nevertheless, small cavities of MOFs usually result in decreased substrate affinity. Therefore, the enzymatic activity of the immobilized enzyme is decreased, when compared to native enzyme activity [15]. Cross-linking immobilization is an alternative, which requires the reaction between cross-linking protein molecules and a chemical cross-linker, usually glutaraldehyde [14]. The diversity of immobilization techniques allows the immobilization of enzymes in distinct materials, such as carbon nanostructures, (carbon black, nanotubes, and graphene and derivatives, among others), ceramic or polymeric matrices, and nanoparticles [16, 17]. It is noteworthy that the performance of an enzymatic biosensor is strongly dependent of the enzyme immobilization, which affects important parameters such as response time, stability, reproducibility, and sensitivity [18].

Another element that interferes in enzymatic biosensor response is active site location. Since proteins are molecules with a giant structure, the active center often can be closed in the molecule's center, making it a very inaccessible site and less susceptible for electron transfer processes. In these cases, a mediator can be used to facilitate the electron transfer between the active site of the enzyme and the modified active electrode. There are several mediators for that, but some are specific for only one enzyme. Regarding Barsan et al. [19], several electrodes modified by functionalized carbon nanotubes act as an alternative to promote the increase of interaction between enzymes and modified electrodes. In addition, they improve the electron transfer rate, besides the fact that phenolic molecules can be used as mediators in these processes. On the other hand, it is also common to use organic dyes such as methylene blue, safranine O, and neutral red [20] and metal complexes, for example, ferrocene [21], as mediators.

2.3 Crude extracts as enzyme sources for biosensing applications

Some enzymes that can be used in phenolic biosensing are peroxidases and polyphenol oxidases. Peroxidases (E.C. 1.11.1) comprise a large family of heme-containing enzymes that react with their substrates using peroxide of hydrogen (H_2O_2) as a proton acceptor, generating water (H_2O) and the oxidized substrate. These enzyme families have been widely used in clinical diagnostics, biosensing, and degradation of pollutants in water [22]. Polyphenol oxidase (E.C. 1.10.3.1) is another enzyme family that includes laccases and tyrosinases, also known as blue-copper oxidases. Laccase enzymes catalyze the oxidation of many phenolic substrates (most commonly ortho- and para-diphenols) with the concomitant reduction of molecular oxygen to water [23], while tyrosinases are enzymes that catalyze two distinct oxygen-dependent subsequent reactions: the hydroxylation of monophenols to ortho-diphenols and the subsequent oxidation of ortho-diphenols to ortho-quinones [24]. These enzymes are very much used in biosensor construction, being often purchased at their active lyophilized form. In the cited cases, the common commercial peroxidases are extracted from Horseradish (*Armoracia rusticana*) roots, while laccases and tyrosinases are extracted from fungi [24].

Oxidoreductases are widely distributed in the plant kingdom, being found in many vegetables. The vegetable crude extracts represent a good alternative to replace manufactured enzymes in biotechnological applications. Commercial enzymes have the advantage of exhibiting high purity levels, which is responsible for a significant increase in selectivity of the analytical device; however, they are very expensive. The crude extracts as enzymatic sources show some advantages such as abundant and easy enzyme obtainment, low cost, and bioavailability of cofactors when necessary to enzymatic activity [25].

Usually, the crude extracts are prepared by processing vegetal tissues in a buffer solution, close to physiological pH, followed by separation of solids by centrifugation. Peroxidases and polyphenol oxidases are found in cell membranes of many vegetables and detergent solutions, such as sodium dodecyl sulfate (SDS), which are dissolved by phenolic compounds to perform the extraction and at the same time that activates the enzyme latent forms [26]. Phenolic compounds are common in vegetables and they react with peroxidases or polyphenol oxidases in the crude extract preparation. In order to preserve enzyme reactivity, phenol scavenger polymers, such as polyvinylpyrrolidones (PVPs) and their derivatives, are added to the extract. These polymers work as phenol adsorbents, interacting with phenolic compounds via hydrogen bonds, preventing these reactions [27].

Several examples of biosensors prepared with crude extracts as enzyme sources were reported [28–30]. Many studies aim to obtain less expensive biosensors with higher durability, since crude extracts mimic the natural enzyme environment. In addition, cofactors and coenzymes can be present in the crude extract. Martins et al. [28] reported the preparation of a biosensor using the crude extract of the pumpkin *Cucubita pepo* for paracetamol detection in aqueous solution, and Benjamin et al. [29] reported a biosensor prepared with a crude extract, which was a source of the polyphenol oxidase, anchored with cerium nanoparticles for rutin detection in solution, showing a limit of detection of the 0.16 μmol L^{-1}.

The biosensor for phenolic compounds from drugs and industrial wastewater was proposed by Antunes et al. [30]. They used the crude extract from vegetal issue sources of polyphenol oxidase, which was anchored on the electrode surface, and the analysis was carried out in an electrochemical cell. The biosensor was evaluated for the quantitative determination of acetaminophen, acetylsalicylic acid, methyldopa, ascorbic acid, and phenolic compounds in a real sample. The limit of detection achieved was 7 μmol of phenol, which is compared to the limit of detection of 8 μmol for polyphenol oxidase for pharmacological samples.

There are several electrochemical biosensors to determine the pharmacological properties of phenolic compounds. Tyrosinase-based biosensor is widely used for detection of phenolic compounds [31, 32]. Its construction is based on the same approaches, such as electropolymerization and sol-gel and polymer entrapment [33]. Aranganathan et al. [33] reported the use of tyrosinase for detection of 3,4-dihydroxy-L-phenylalanine (L-DOPA), which is a preferred drug for the treatment of Parkinson's disease. Florescu and David [34] developed a tyrosinase-based biosensor for selective dopamine detection, in which its selectivity was increased by employing cobalt (II)-porphyrin (CoP) film-modified gold electrodes. It operates by enabling the direct immobilization of the enzyme layer in more available sites, acting as an electrochemical mediator during enzyme-catalyzed reaction, leading to a complete recovery of the electrode, with no effect on the detection limit [34]. Tyrosinase can be used as a pesticide detector as well. In this respect, Liu et al. [35] developed a biosensor consisting of a glassy carbon electrode modified with graphene and containing tyrosinase immobilized on platinum nanoparticles. It was for organophosphorus pesticide detection and they found that the presence of Pt nanoparticles and graphene improved the biosensor sensitivity by enhancing the efficiency of the electrochemical reduction of o-quinone. Also, in the study conducted by Everett and Rechnitz [36], the tyrosinase-based biosensor was very sensitive to pesticide in aqueous solution.

Peroxidase-based biosensors are alternatives to determine phenol and phenolic compounds. The need for a peroxidase-based material that would be more stable in aqueous media, with lower costs, leads to the use of hemoglobin in the biosensor processing [31]. Highly sensitive hemoglobin-based biosensor was obtained by the modification of a carbon-paste electrode with hemoglobin and multiwalled carbon nanotubes. It was tested in the detection of methylparaben, present in real samples of urine and human serum. It reached a detection limit of 25 nM [37]. In their work, Haijan et al. [37] also showed that the immobilization of hemoglobin onto cuprous sulfide nanorods/Nafion® nanocomposite film is an effective way to construct a biosensor for polyphenol detection. In addition to the hemoglobin immobilization, the polyphenol detection was also enhanced.

Rodríguez-Delgado et al. [23] developed laccase-based biosensors that presented high sensitivity and reproducibility for phenolic compounds *in situ* and environmental monitoring. Several others pollutants, that can be easily dissolved in water and, therefore, are considered environmental pollutants, must be monitored. It is the case of several compounds used by the food and textile industries. With this regard, tartrazine, a synthetic organic food azo dye, has its use controlled due to its potential harmfulness to human health. The first work on the use of laccase-based biosensor for the determination of tartrazine dye was recently developed by Mazlan et al. [38], which is a biosensor consisting of laccase enzyme immobilized on methacrylate-acrylate microspheres and composites with gold nanoparticles [38].

The adverse use of drugs based on morphine and narcotics causes several illnesses around the world. The development of efficient methods to detect illicit drugs in biological samples, such as urine and blood plasma, is, therefore, much required. Gandhi et al. [39] reported the advances in the field of biosensors for narcotic drug detection. Among them, they showed that the double-stranded DNA (ds-DNA) immobilized onto mercaptobenzaldehyde-modified Au electrode is an advantageous and promising biosensor to morphine detection, since it presents the advantage of no need of additional steps of extraction, cleansing, and derivatization [40].

Regarding drug detections, yet Alvau et al. [41] proposed a biosensor for therapeutic drug monitoring based on acetylcholinesterase (AChE) and choline oxidase. These are promising biosensors because they also present the possibility of distinct application, for instance, AChE-based biosensors can find application in environmental monitoring, since they can be used for the electrochemical detection of organophosphate and carbamate pesticides. The global concern over pesticide level increase rose the last decade due to the high toxicity and bioaccumulation effects of such compounds, and the significant risks that they represent to the environment and human health. Therefore, monitoring pesticide residues by sensitive analytical techniques is indispensable. In view of the harmful effects associated with pesticides, a legislative framework has been established worldwide which defines rules for the approval of active chemicals and maximum residue levels (MRLs) allowed in food and water. The legal limits for the amount of pesticides allowed in food and drinking water are set by the Environmental Protection Agency (EPA) in USA and for the European Environment Agency in European Union (EEA). These government agencies establish the appropriate pesticides levels, according to the type of crops and substance. For instance, the pesticide methomyl has the maximum tolerance established at 2.0 ppm (parts per million) in lemon in USA, whereas EEA established a MRL lower than 0.01 ppm for the same pesticide in lemon. However, in the case of the pesticide chlorpyrifos in apples, both agencies authorize the same MRL in 0.01 ppm for apples. Commonly, the MRLs are in the range of ppm to ppb (parts per billion); nevertheless, there are some pesticides that are forbidden and are illegally used. In contrast, in Brazil, the legislation regarding the use of pesticides in crops as well as the detection limit in food and water is much more permissive. For instance, it allows a level of glyphosate in water up to 5000 times greater than that allowed in the European Union. Over the years, several biomolecules have been used as a biorecognition element in biosensors for pesticide detection, such as cells, antibodies, aptamers, and enzymes. In this section, we will focus on enzymatic biosensors for organophosphate (OP) and carbamate quantification based on electrochemical transducer. These devices use acetylcholinesterase (AChE) and butyrylcholinesterase (BChE), in addition to alkaline phosphatase (ALP) and organophosphorus hydrolase (OPH) for OP detection, specifically. AChE-based biosensors are among the most popular electrochemical sensing platforms for the aforementioned types of pesticides [42]. AChE is susceptible to be inhibited by OPs as well as carbamate pesticides. The working mechanism of an electrochemical AChE-based biosensor is based on inhibitory effects. In the absence of OPs and carbamates (analytes), the substrate acetylthiocholine is converted into thiocholine and acetate. Afterwards, thiocholine is oxidized by the applied potential. When the analyte is present in the solution, AChE has its activity decreased by the pesticide inhibition. Consequently, the conversion of acetylthiocholine is partial or totally reduced, and the pesticides are indirectly detected [43]. **Figure 1** shows the working principle of AChE biosensor.

Selectivity is the most significant hallmark of enzymatic biosensors. In the case of AChE-based biosensors, it is only possible to detect an assortment of pesticides in a complex matrix, and no qualitative or quantitative information is obtained for a single inhibitor. Besides, AChE can be inhibited by heavy metals, drugs, and nerve agents. Therefore, the inhibition strategy to detect pesticides towards AChE implies in poor selectivity [44]. An important consideration is that AChE inhibition by pesticides may diverge according to the source of enzyme. Studies have demonstrated that AChE extracted from electric eel exhibited greater sensitivity in comparison to those from bovine and human erythrocytes [45]. On the other hand, genetically modified AChE from *Drosophila melanogaster* revealed superior results [45]. In order to address these limitations, numerous approaches have been developed, involving nanomaterial technologies to improve the transducer performance in addition to genetic engineering [46].

The design of novel AChE-based biosensors for pesticide detection concerns the application of nanomaterials offering transducing platforms with outstanding electrochemical behavior. The advantages provided by nanomaterials in electrochemical sensing are associated with large surface-to-volume ratio, controlled morphology, electrocatalytic properties, immobilization of biomolecules, and possibilities of system miniaturization [47].

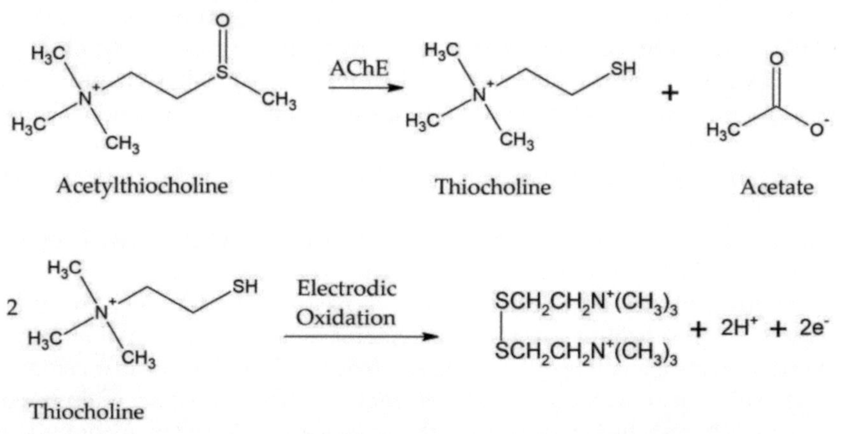

Figure 1.
Scheme of the general reaction mechanism of an electrochemical biosensor based on AchE.

Currently, the employment of screen printed electrodes (SPEs) has boosted the scenario of AChE-based biosensors. Those electrodes promote the system miniaturization addressing the sample volume issues, combining cost effectiveness and simple manipulation. Therefore, several strategies of modification have been applied to achieve high sensitivity and low limit of detection. A smart AChE biosensor approach used homemade SPE modified with single-walled carbon nanotubes (SWCNT) derivatized with cobalt phthalocyanine to detect thiocholine at a lower overpotential in comparison to bare SPE and SPE modified with nonfunctionalized SWCNT in only 80 μL of sample [48]. Remarkably, the performance of an AChE-based biosensor was improved due to electrode modification with *N*-carbamoylmaleimide-functionalized carbon dots (*N*-MAL-CDs) as a nanostabilizer [49]. The initial electrochemical signals of thiocholine were obtained without signal loss, as a result of the Michael addition reaction functionalizing CDs with *N*-MAL. Then, *N*-MAL-CDs can react with thiol group from thiocholine, forming a thiol containing compound. The aforementioned compound cannot be easily oxidized during the detection process, avoiding the signal loss. For the fabrication of AChE/*N*-MAL-CDs/SPE biosensor, they used a commercial SPE in which all electrochemical measurements were performed in a droplet of 50 μL. One significant breakthrough offered by SPE is the simultaneous analysis performed by an array of electrodes [50]. The multiplexed analysis integrated into an automated system enables the rapid detection of OP pesticides being convenient for commercial and routine applications. Hence, an array with 12 SPEs deposited in sequence side by side on a ceramic substrate in which the working electrode was printed with a carbon ink containing cobalt phthalocyanine and $Ag/AgCl/KCl_{sat}$ was used as reference/counter electrode. By means of using six types of recombinant AChE, it was possible to acquire qualitative and quantitative information through inhibition assay since the enzyme becomes selective among the OP pesticides, such as dichlorvos, malaoxon, chlorpyrifos-oxon, chlorpyrifos-methyl-oxon, chlorfenvinphos, and pirimiphos-methyl-oxon.

Despite all exceptional SPE properties, they present certain drawbacks, such as the dissolution of conductive and insulating inks due to use of organic solvents, lack of reproducibility, and need of pretreatment procedure.

The continuous progress in biosensing area leads to the development of paper-based analytical devices (PADs) with electrochemical detection. The PADs have emerged as a powerful analytical tool integrating the convenience of SPEs, i.e., portability, simplicity with easy manufacturing of paper, availability, and reduced cost. Furthermore, the PADs provide singular advantages since they can be scalable manufactured from renewable sources, biocompatibility, biodegradable, and low cost. A pioneering research involving a paper-based amperometric sensor for AChE determination was based on screen printed graphene electrodes fabricated by a wax printing method to obtain the detection area. The approach was applied for blood sample analysis, but it has potential to be used for pesticide detection.

Numerous immobilization strategies and fabrication methods have brought new perspectives to AChE-based biosensors. The investigations have focused on enzyme stability, reproducibility, miniaturization, and mass production [51]. The usage of smartphones in biosensing has played new horizons in environmental monitoring; however, it remains a challenge [52]. The electrochemical biosensors on smartphone use portable electrical detectors for amperometric, potentiometric, and impedimetric measurements, but environmental analyses are still scarce. Although great progress has been made with wireless biosensors, there is a lack of applications in pesticide detection.

3. Conclusions

In this chapter, we presented the state of the art of biosensors to detect phenolic compounds, with environmental and pharmacological applications. Monitoring the negative environmental impacts of phenolic compounds uses has attracted researchers' attention since these compounds are widely applied in several industrial sectors. Among the biosensors developed for environmental monitoring, enzymatic ones are the most prominent used for phenolic compound detection. By combining enzymes with electrochemical transducers, cheaper devices had been developed, which is a great advantage to environmental analytical methods. The immobilization of enzymes on the electrode surface consists in physical and chemicals interactions. The location of actives sites is important to biosensor response; nevertheless, mediators can be used to transpose this barrier and facilitate the electronic transfer needed for the detection process. Tyrosinase-based biosensor is the most common biosensor for phenolic compound detection, which is a precursor for drugs for the treatment of Parkinson's disease, as morphine-based drugs. Also, acetylcholinesterase-based biosensors are widely employed because they present high efficiency to detect organophosphate and carbamate compounds, which are used as pesticides. The design of novel AChE-based biosensors for pesticide detection concerns the application of nanomaterials offering transducing platforms with outstanding electrochemical behavior. The employment of screen printed electrodes promotes the system miniaturization, which is a new perspective to electrochemical biosensor application.

Acknowledgements

The authors thank Conselho Nacional de Desenvolvimento Científico e Tecnológico (CNPq), grant number 554569/ 2010-8, and Coordenação de Aperfeiçoamento de Pessoal de Nível Superior (CAPES) for financial support and scholarships, grant number 88882.306480/2018-01.

Conflict of interest

The authors state that there is no conflict of interest associated with this work.

Author details

Flavio Colmati*, Lívia Flório Sgobbi, Guilhermina Ferreira Teixeira,
Ramon Silva Vilela, Tatiana Duque Martins and Giovanna Oliveira Figueiredo
Chemistry Institute, Federal University of Goiás, Goiânia, GO, Brazil

*Address all correspondence to: colmati@ufg.br

References

[1] Korotcenkov G, Han SD, Stetter JR. Review of electrochemical hydrogen sensors. Chemical Reviews. 2009;**109**(3):1402-1433. DOI: 10.1021/cr800339k

[2] Pandey VP, Awasthi M, Singh S, Tiwari S, Dwivedi UN. A comprehensive review of function and application of plant peroxidase. Biochemistry & Analytical Biochemistry. 2017;**6**(1):308-324. DOI: 10.4172/2161-1009.1000308

[3] Nigam VK, Shukla P. Enzyme based biosensors for detection of environmental pollutants–A review. Journal of Microbiology and Biotechnology. 2015;**25**(11):1773-1781. DOI: 10.4014/jmb.1504.04010

[4] Vermerris W, Nicholson R, editors. Phenolic Compound Biochemistry. Dordrecht: Springer Netherlands; 2006. 276 p. DOI: 10.1007/978-1-4020-5164-7

[5] Wells MJM, Fono LJ, Pellegrin M-L, Morse A. Emerging pollutants. Water Environment Research. 2007;**79**(10):2192-2209. DOI: 10.2175/106143007x218719

[6] Nadal A, Quesada I, Tudurí E, Nogueiras R, Alonso-Magdalena P. Endocrine-disrupting chemicals and the regulation of energy balance. Nature Reviews Endocrinology. 2017;**13**:536. DOI: 10.1038/nrendo.2017.51

[7] Damstra TBS, Bergman A, Kavlock R, Van der Kraak G. International Program on Chemical Safety. Global Assessment of Endocrine-Disrupting Chemicals. [Internet]. 2002. Available from: http://www.who.int/ipcs/publications/en/ch2.pdf?ua=1 [Accessed: December 06, 2018]

[8] Bouchard P. Endocrine-disrupting chemicals, a multifaceted danger. Comptes Rendus Biologies. 2017;**340**(9):401-402. DOI: 10.1016/j.crvi.2017.09.001

[9] Di Fusco M, Tortolini C, Deriu D, Mazzei F. Laccase-based biosensor for the determination of polyphenol index in wine. Talanta. 2010;**81**(1):235-240. DOI: 10.1016/j.talanta.2009.11.063

[10] Wang J. Analytical Electrochemistry. 3rd ed. Danvers: Wiley-VCH; 2006. 272 p. DOI: 10.1002/0471790303

[11] Sadeghi SJ. Amperometric biosensors. In: Roberts GCK, editor. Encyclopedia of Biophysics. Berlin, Heidelberg: Springer Berlin Heidelberg; 2013. 67 p. DOI: 10.1007/978-3-642-16712-6_713

[12] Dzyadevych S, Jaffrezic-Renault N. 6—Conductometric biosensors. In: Schaudies RP, editor. Biological Identification. Elsevier Ltd, Woodhead Publishing; 2014. 153-193 p. DOI: 10.1533/9780857099167.2.153

[13] Mataveli LV, Antunes NJ, Brigagão MRPL, Magalhães CS, Wisniewski C, Luccas PO. Evaluation of a simple and low cost potentiometric biosensor for pharmaceutical and in vivo adrenaline determination. Biosensors & Bioelectronics. 2010;**26**(2):798-802. DOI: 10.1016/j.bios.2010.06.021

[14] Pereira AR, Sedenho GC, Souza JCP, Crespilho FN. Advances in enzyme bioelectrochemistry. Anais da Academia Brasileira de Ciências. 2018;**90**(1):825-857. DOI: 10.1590/0001-3765201820170514

[15] Zhang C, Wang XR, Hou M, Li XY, Wu XL, Ge J. Immobilization on metal-organic framework engenders high sensitivity for enzymatic electrochemical detection. ACS Applied Materials & Interfaces.

2017;**9**(16):13831-13836. DOI: 10.1021/acsami.7b02803

[16] Andreescu S, Njagi J, Ispas C. Nanostructured materials for enzyme immobilization and biosensors. In: Erokhin V, Ram MK, Yavuz O, editors. The New Frontiers of Organic and Composite Nanotechnology. Amsterdam: Elsevier; 2008. pp. 355-394. DOI: 10.1016/B978-008045052-0.50009-9

[17] Ansari SA, Husain Q. Potential applications of enzymes immobilized on/in nano materials: A review. Biotechnology Advances. 2012;**30**(3):512-523. DOI: 10.1016/j.biotechadv.2011.09.005

[18] Sassolas A, Blum LJ, Leca-Bouvier BD. Immobilization strategies to develop enzymatic biosensors. Biotechnology Advances. 2012;**30**(3):489-511. DOI: 10.1016/j.biotechadv.2011.09.003

[19] Barsan MM, Ghica ME, Brett CMA. Electrochemical sensors and biosensors based on redox polymer/carbon nanotube modified electrodes: A review. Analytica Chimica Acta. 2015;**881**:1-23. DOI: 10.1016/j.aca.2015.02.059

[20] Salomi BSB, Mitra CK. Electrochemical studies on horseradish peroxidase covalently coupled with redox dyes. Biosensors & Bioelectronics. 2007;**22**(8):1825-1829. DOI: 10.1016/j.bios.2006.08.042

[21] Sadeghi SJ, Gilardi G, Cass AEG. Mediated electrochemistry of peroxidases—Effects of variations in protein and mediator structures. Biosensors & Bioelectronics. 1997;**12**(12):1191-1198. DOI: 10.1016/S0956-5663(97)00089-4

[22] Regalado C, García-Almendárez BE, Duarte-Vázquez MA. Biotechnological applications of peroxidases. Phytochemistry Reviews. 2004;**3**(1):243-256. DOI: 10.1023/b:phyt.0000047797.81958.69

[23] Rodríguez-Delgado MM, Alemán-Nava GS, Rodríguez-Delgado JM, Dieck-Assad G, Martínez-Chapa SO, Barceló D, et al. Laccase-based biosensors for detection of phenolic compounds. TrAC, Trends in Analytical Chemistry. 2015;**74**:21-45. DOI: 10.1016/j.trac.2015.05.008

[24] Vicentini FC, Janegitz BC, Brett CMA, Fatibello-Filho O. Tyrosinase biosensor based on a glassy carbon electrode modified with multi-walled carbon nanotubes and 1-butyl-3-methylimidazolium chloride within a dihexadecylphosphate film. Sensors and Actuators B: Chemical. 2013;**188**:1101-1108. DOI: 10.1016/j.snb.2013.07.109

[25] Fatibello-Filho O, Vieira I. Uso analítico de tecidos e de extratos brutos vegetais como fonte enzimática. Quimica Nova. 2002;**25**(3):455-464. DOI: 10.1590/S0100-40422002000300019

[26] Gouzi H, Moreau T, Depagne C, Coradin T. Immobilization of a polyphenol oxidase extract from Terfezia leonis Tul. Desert truffle in multilayered silica films for dopamine biosensing. Silicon. 2013;**5**(4):241-246. DOI: 10.1007/s12633-013-9165-z

[27] Hatti-Kaul R, Mattiasson B. Release of protein from biological host. In: Hatti-Kaul R, Mattiasson B, editors. Isolation and Purification of Proteins. New York: Marcel Dekker, Inc.; 2003. 696 p

[28] Martins TD, Ribeiro ACC, Colmati F, Souza GA, Camargo HS, Dias DL, et al. Supramolecular materials for optical and electrochemical biosensors. In: Rinken T, editor. Biosensors. London: IntechOpen; 2015. 138-164 p. DOI: 10.5772/60710

[29] Benjamin SR, Vilela RS, de Camargo HS, Guedes MIF, Fernandes KF,

Colmati F. Enzymatic electrochemical biosensor based on multiwall carbon nanotubes and cerium dioxide nanoparticles for rutin detection. International Journal of Electrochemical Science. 2018;**13**(1):563-586. DOI: 10.20964/2018.01.51

[30] Antunes R, Ferraz D, Garcia L, Thomaz D, Luque R, Lobón G, et al. Development of a polyphenol oxidase biosensor from jenipapo fruit extract (*Genipa americana* L.) and determination of phenolic compounds in textile industrial effluents. Biosensors. 2018;**8**(2):47. DOI: 10.3390/bios8020047

[31] Kafi AKM, Lee D-Y, Park S-H, Kwon Y-S. Potential application of hemoglobin as an alternative to peroxidase in a phenol biosensor. Thin Solid Films. 2008;**516**(9):2816-2821. DOI: 10.1016/j.tsf.2007.04.123

[32] Karim F, Fakhruddin ANM. Recent advances in the development of biosensor for phenol: A review. Reviews in Environmental Science and Bio/Technology. 2012;**11**(3):261-274. DOI: 10.1007/s11157-012-9268-9

[33] Aranganathan V, Kavya C, Sheeja MS, Yuvraj C. Construction of biosensor for detection of phenolic compound using thermostabilized agaricus bisporus tyrosinase. Arabian Journal for Science and Engineering. 2017;**42**(1):11-18. DOI: 10.1007/s13369-016-2044-3

[34] Florescu M, David M. Tyrosinase-based biosensors for selective dopamine detection. Sensors. 2017;**17**(6):1314-1330. DOI: 10.3390/s17061314

[35] Liu T, Xu M, Yin H, Ai S, Qu X, Zong S. A glassy carbon electrode modified with graphene and tyrosinase immobilized on platinum nanoparticles for sensing organophosphorus pesticides. Microchimica Acta. 2011;**175**(1):129-135. DOI: 10.1007/s00604-011-0665-5

[36] Everett WR, Rechnitz GA. Mediated bioelectrocatalytic determination of organophosphorus pesticides with a tyrosinase-based oxygen biosensor. Analytical Chemistry. 1998;**70**(4):807-810. DOI: 10.1021/ac970958l

[37] Hajian A, Ghodsi J, Afraz A, Yurchenko O, Urban G. Nanomolar detection of methylparaben by a cost-effective hemoglobin-based biosensor. Materials Science and Engineering: C. 2016;**69**:122-127. DOI: 10.1016/j.msec.2016.06.061

[38] Mazlan SZ, Lee YH, Hanifah SA. A new laccase based biosensor for tartrazine. Sensors. 2017;**17**(12):2859. DOI: 10.3390/s17122859

[39] Gandhi S, Suman P, Kumar A, Sharma P, Capalash N, Suri CR. Recent advances in immunosensor for narcotic drug detection. BioImpacts: BI. 2015;**5**(4):207-213. DOI: 10.15171/bi.2015.30

[40] Talemi RP, Mashhadizadeh MH. A novel morphine electrochemical biosensor based on intercalative and electrostatic interaction of morphine with double strand DNA immobilized onto a modified Au electrode. Talanta. 2015;**131**:460-466. DOI: 10.1016/j.talanta.2014.08.009

[41] Alvau MD, Tartaggia S, Meneghello A, Casetta B, Calia G, Serra PA, et al. Enzyme-based electrochemical biosensor for therapeutic drug monitoring of anticancer drug irinotecan. Analytical Chemistry. 2018;**90**(10):6012-6019. DOI: 10.1021/acs.analchem.7b04357

[42] Van Dyk JS, Pletschke B. Review on the use of enzymes for the detection of organochlorine, organophosphate and carbamate pesticides in the environment. Chemosphere. 2011;**82**(3):291-307. DOI: 10.1016/j.chemosphere.2010.10.033

[43] Songa EA, Okonkwo JO. Recent approaches to improving selectivity and sensitivity of enzyme-based biosensors for organophosphorus pesticides: A review. Talanta. 2016;**155**:289-304. DOI: 10.1016/j.talanta.2016.04.046

[44] Luque de Castro MD, Herrera MC. Enzyme inhibition-based biosensors and biosensing systems: Questionable analytical devices. Biosensors & Bioelectronics. 2003;**18**(2-3):279-294. DOI: 10.1016/S0956-5663(02)00175-6

[45] de Oliveira Marques PRB, Nunes GS, dos Santos TCR, Andreescu S, Marty J-L. Comparative investigation between acetylcholinesterase obtained from commercial sources and genetically modified drosophila melanogaster: Application in amperometric biosensors for methamidophos pesticide detection. Biosensors and Bioelectronics. 2004;**20**(4):825-832. DOI: 10.1016/j.bios.2004.03.021

[46] Pundir CS, Chauhan N. Acetylcholinesterase inhibition-based biosensors for pesticide determination: A review. Analytical Biochemistry. 2012;**429**(1):19-31. DOI: 10.1016/j.ab.2012.06.025

[47] Kurbanoglu S, Ozkan SA, Merkoçi A. Nanomaterials-based enzyme electrochemical biosensors operating through inhibition for biosensing applications. Biosensors & Bioelectronics. 2017;**89**:886-898. DOI: 10.1016/j.bios.2016.09.102

[48] Jubete E, Żelechowska K, Loaiza OA, Lamas PJ, Ochoteco E, Farmer KD, et al. Derivatization of SWCNTs with cobalt phthalocyanine residues and applications in screen printed electrodes for electrochemical detection of thiocholine. Electrochimica Acta. 2011;**56**(11):3988-3995. DOI: 10.1016/j.electacta.2011.01.123

[49] Xu J, Yu C, Feng T, Liu M, Li F, Wang Y, et al. N-Carbamoylmaleimide-treated carbon dots: Stabilizing the electrochemical intermediate and extending it for the ultrasensitive detection of organophosphate pesticides. Nanoscale. 2018;**10**(41):19390-19398. DOI: 10.1039/c8nr05098h

[50] Crew A, Lonsdale D, Byrd N, Pittson R, Hart JP. A screen-printed, amperometric biosensor array incorporated into a novel automated system for the simultaneous determination of organophosphate pesticides. Biosensors & Bioelectronics. 2011;**26**(6):2847-2851. DOI: 10.1016/j.bios.2010.11.018

[51] Uniyal S, Sharma RK. Technological advancement in electrochemical biosensor based detection of organophosphate pesticide chlorpyrifos in the environment: A review of status and prospects. Biosensors & Bioelectronics. 2018;**116**:37-50. DOI: 10.1016/j.bios.2018.05.039

[52] Zhang D, Liu Q. Biosensors and bioelectronics on smartphone for portable biochemical detection. Biosensors & Bioelectronics. 2016;**75**:273-284. DOI: 10.1016/j.bios.2015.08.037

Chapter 5

Challenges and Applications of Impedance-Based Biosensors in Water Analysis

Kairi Kivirand, Mart Min and Toonika Rinken

Abstract

Monitoring of the environment is a global priority due to the close connection between the environmental pollution and human health. Many analytical techniques using various methods have been developed to detect and monitor the levels of pollutants (pesticides, toxins, bacteria, drug residues, etc.) in natural water bodies. The latest trend in modern analysis is to measure pollutants in real-time in the field. For this purpose, biosensors have been employed as cost-effective and fast analytical techniques. Among biosensors, impedance biosensors have significant potential for use as simple and portable devices. These sensors involve application of a small amplitude AC voltage to the sensor electrode and measurement of the in-/out-of-phase current response as a function of frequency integrated with some biorecognition element on the sensing electrodes that can bind to the target, modifying the sensor electrical parameters. However, there are some drawbacks concerning their selectivity, stability, and reproducibility. The aim of this paper is to give a critical overview of literature published during the last decade based on the development issues of impedimetric biosensors and their applicability in water analysis.

Keywords: electrochemical impedance spectroscopy, biosensor, challenges, application, water analysis

1. Introduction

Pollution of water by different chemicals disturbs ecosystems. Pollutants can also accumulate in the environment and can be found for many years after they have been banned. In addition, pollutants may accumulate into our food chain (seafood, drinking water, agricultural products, etc.) and thereby affect all living organisms including humans [1]. Some pollutants can be found years after having been banned. For example, despite being banned for agricultural use in EU in 2003 because of ubiquitous and unpreventable water contamination [2], atrazine was even after 5 years still found in spring and groundwaters at quantities between 0.9 and 2.8% of the annually applied amount before the ban [3]. Therefore, monitoring of natural water has become an essential requirement worldwide. Currently, the most common option to detect pollution is the use of fixed monitoring stations, which need trained people to analyze the collected data and are usually quite expensive. To decrease costs and make monitoring more effective, there has been an increasing interest in the development of portable and user-friendly systems, which could give us fast, precise, and reliable information.

Biosensors can be a useful tool for the detection of pollutants in the water. In comparison with traditional monitoring techniques, biosensors are portable, need minimal sample preparation, and are also rapid and reliable [4]. According to the International Union of Pure and Applied Chemistry (IUPAC) definition, a biosensor is a self-contained, integrated receptor transducer device, which is capable of providing selective quantitative or semiquantitative analytical information and which uses a biological recognition element (bio-receptor) and a transducer in direct special contact [5]. Biosensors can be used for continuous monitoring with high selectivity and sensitivity.

Biosensors are classified according to their biorecognition element or signal transducer into various categories. Electrochemical biosensors based on impedance are among the most promising ones due to their portability, rapidity, and label-free operation. Label-free sensors register changes in the electrical properties due to interactions between biological molecule attached to the sensor and an analyte present in the sample, and as these sensors generate rapid response, they can be used to track molecular events in a real-time manner [6]. The main advantage of label-free detection is that it is possible to acquire direct information of the interactions between native proteins and ligands [6, 7]. In environmental analysis most of the biosensors used are enzyme-based biosensors [8–12] or antibody-based immunosensors [13–16]. In recent years also the development of aptasensors has increased [17–19]. The present chapter gives a critical overview of the development issues and applicability of different impedimetric biosensors used for water analysis.

2. Electrochemical impedance spectroscopy

Electrochemical impedance spectroscopy (EIS) is an analytical tool, which has been used for studying electrochemical systems including corrosion [20–22], battery development [23], electrodeposition [24], fuel cells [25, 26], and charge transport through membranes [27]. For impedance measurements, the alternating current (AC) voltage applied is typically small (up to 10 mV) so that the voltage-current response is linear, allowing simple equivalent circuit analysis [28]. Different waveforms of the AC voltage $V(t)$ varying in time can be used [29]. The simplest but best-known waveform among them is a pure sine wave $V(t) = V_0 \sin(\omega t)$, which varies periodically (oscillates) with angular frequency $\omega = 2\pi f$, rad/s, where f, ($1/s \equiv Hz$), is the repetition frequency of oscillation periods. The current response $I(t)$ to the applied voltage $V(t)$ is also the sine wave at exactly the same frequency $\omega = 2\pi f$. In addition, the current response $I(t)$ is shifted over the time interval (Δ t) against the applied voltage $V(t)$ because of containing inert energy saving components (capacitance C and/or inductance L) of impedance Z. In practice, it is reasonable to use the phase shift $\varphi = 2\pi f(\Delta t)$, rad, instead of the time interval (Δt). Predominantly, the impedance handling assumes that there are no changes in impedance value during the observation time interval. Therefore, we can exclude time dependence from the mathematical expression of impedance and use the frequency dependent impedance $Z(\omega)$ instead of $Z(t,\omega)$. Mathematical equation for the impedance $Z(\omega)$ is the ratio between the voltage-time function $V(t)$ and the resulting current-time function $I(t)$ (Eq. (1)):

$$Z(t) = \frac{V(t)}{I(t)} = \frac{V_0 \sin(2\pi ft)}{I_0 \sin(2\pi ft + \varphi)} = \frac{V_0 \sin(\omega t)}{I_0 \sin(\omega t + \varphi)} \quad (1)$$

More complicated voltage signal waveforms are required for the fast performance of EIS by generating the signal components at several frequencies simultaneously [29]. As EIS measures the response of an electrochemical cell to a voltage at different frequencies, the data obtained allows characterizing the complex electrode systems on layers, surfaces, or membranes where electrical charge transfer and ion diffusion processes take place [7]. To evaluate and interpret the results, the EIS data are usually analyzed using Bode or Nyquist plots [30, 31].

Based on the methodologies of signal collection, impedimetric detection can be categorized in two ways: capacitive faradaic or non-faradaic. It is important to distinguish between those approaches. In electrochemical terminology, a faradaic process is the one where charge is transferred across an interface. In the case of non-faradaic, the transient currents can flow without charge transfer (e.g., charging a capacitor). In faradaic EIS, a redox probe is alternately oxidized and reduced by the transfer of an electron to and from the metal electrode. Thus, faradaic EIS requires the addition of a redox probe and direct current (DC) bias conditions such that it is not depleted. In contrast, no additional reagent is required for non-faradaic impedance spectroscopy, rendering non-faradaic schemes somewhat more amenable to point-of-care applications [32, 33].

In the case of faradaic impedimetry, the electrode surface is partially or fully covered with a non-isolating layer or with an isolating layer able to catalyze a redox probe [34]. Non-faradaic approach is also known as the direct measurement manner (without chemical reactions). In the case the redox probe is missing, the impedance depends on the conductivity of the supporting electrolyte and electrode interfacial properties. Capacitive approach means that the surface of the electrode is completely covered with a dielectric layer. In this type of sensors, no redox probe is present in the system; and the current is measured under a small amplitude sinusoidal voltage signal, at low frequencies [34]. Capacitive biosensors are mainly based on a non-faradaic approach, because the transient current flows without charge transfer and no additional reagent is required.

Briefly in faradaic approach, the charge is transferred across the electrified interface as a result of an electrochemical reaction, and in non-faradaic approach, the charge is associated with movement of electrolyte ions, reorientation of solvent dipoles, adsorption/desorption, etc. at the electrode-electrolyte interface. Detailed overviews about faradic and non-faradaic systems are given in Refs. [31, 34, 35].

In order to present information about surfaces, layers, or membranes after the immobilization of biomolecules, EIS experimental data is often analyzed using an equivalent circuit of electrochemical cell [30]. The Randle's circuit (**Figure 1**) is a frequently used equivalent for modeling the impedance [32]. The non-faradaic sensor comprises the uncompensated resistance of the electrolyte (R_s) and the constant

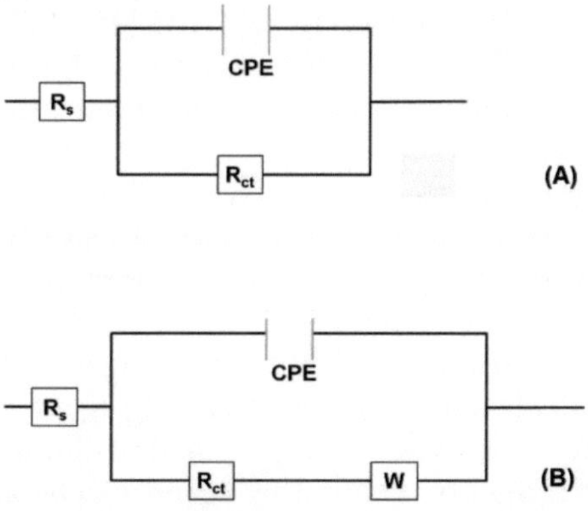

Figure 1.
Simplified circuit models for (A) non-faradaic and (B) faradaic systems. Abbreviations: R_s, resistance of the electrolyte; CPE, constant phase element; R_{ct}, charge-transfer resistance; W, the Warburg impedance.

phase element (CPE) having capacitive-like properties in parallel with the charge-transfer resistance (R_{ct}).

Sometimes simplifications are introduced, and the CPE is replaced by a double-layer capacitance (C_{dl}), which introduces the constant phase shift of -π/2 rad (−90°) at all the frequencies. In reality, the CPE introduces the phase shift φ less than π/2 [29, 36].

The faradaic sensor model includes the Warburg impedance (*W*), which describes diffusion phenomenon taking place due to chemical redox processes. The ideal Warburg impedance introduces the phase shift of π/4. Values of the charge transfer R_{ct} and *W* depend on physicochemical parameters of a system. In real systems, impedance spectra are usually more complicated, and, therefore, the Randle's circuit with a corresponding plot may not give proper results [31].

3. Challenges of EIS-based biosensors

The detection of contaminates in water is very important since high pollution (heavy metals, pesticide and antibiotic residues, etc.) or the presence of pathogens (infectious microorganisms like viruses, bacteria, and fungi) can seriously endanger human health.

Several technical challenges hinder the development and construction of EIS-based biosensors: limitations to detect small molecules, reusability, and sufficient stability for repetitive measurements [37, 38]. However, the most crucial problem is whether the impedance biosensors have sufficient selectivity for their application in real samples, which typically contain an unknown amount of nontarget molecules.

There are two main types of impedimetric biosensors—with or without a specific biorecognition element [30]. The most common biorecognition elements used are specific antibodies [39, 40]. The key to the successful performance of EIS-based biosensors is how to decrease the *non-specific bindings* and increase the *selectivity*. Selectivity is particularly important in real samples where the analyte concentration can be much smaller than the concentration of nontarget molecules. Non-specific binding is typically ascribed to proteins contained in a sample matrix attaching to the sensor interface through an unwanted process not involving the bimolecular recognition [41]. One option to decrease non-specific binding is to use blocking agents like bovine serum albumin (BSA), cysteine, or ethanolamine [42–44]. The choice of a blocking agent depends on the particular system. For example, Puttharugsa and Kamolpach used BSA for prevention of non-specific binding on gold electrodes, and the selectivity of the constructed biosensors was testes toward *Escherichia coli* K12 (*E. coli* K12) as a model with EIS [45]. When BSA is adsorbed physically onto the surface, the penetration of redox probe was reduced resulting in the increase of the semicircle Nyquist curve proving that BSA prevents the adsorption of bacteria onto the blocked surface by delaying the interfacial electron-transfer kinetics and increasing the electron-transfer resistance. Riquelme et al. studied several blocking agents (mercaptoundecanol, polyethylene glycol, BSA, and chicken serum albumin) to study the effect of biomolecule size and hydrophilic properties on blocking capacity on gold electrodes [43]. Higher impedance change was observed with lower molecular weight blocking agents, due to higher molecular packing on gold electrode. In addition lower blocking agent concentrations may be required if the electrode surface has already been bio-functionalized.

In addition to blocking agents, antifouling agents can be used to prevent target depletion via non-specific bindings [46]. Blocking agent reduces the non-specific binding by blocking the active functional groups on the surface and can stabilize the biomolecule bound to the surface [47]. Antifouling (or non-fouling) agent is a compound that has the capability to ensure the resistivity to the non-specific adsorption of proteins, cells, or other biological species [48]. Ortiz-Aguayo and Valle tried to decrease the non-specific adsorption to the graphite-epoxy composite electrode surface of an EIS-based wine aptasensor using polyethylene glycol [46]. Even though the aptasensor showed good sensitivity, the blocking did not work so efficiently; and the recovery was approximately 77%.

One of the main challenges is the *sensitivity*, which depends on the thickness of the sensing layer [49–53]. If the sensing layer is too thin, the electrode surface may be exposed, which would decrease the signal to noise ratio and decrease the sensitivity. If the sensing layer is too thick, the detected AC impedance current reduces meaning that the electron transfer between layers is hindered and the sensitivity is decreased. For example, Groß et al. studied the effect of the thickness on the base resistance in the range 30 to 150 μm and found that the sensitivity decreased along with the sensitive layer getting thicker [49]. They also found that there is a trade-off between wide linear range and high sensitivity. In addition, the sensor signal became slower as the thickness of the sensitive layer increased [49]. Functionalization of the electrodes with high-affinity biomolecules enhances besides selectivity and also the sensitivity of the system. Therefore, EIS is very often combined with different nanostructured interfaces in order to increase the amount of biorecognition material on the surface and therefore to improve the sensor sensitivity and extend its linear working [54–59]. This improvement is associated with the dimensions of nanomaterials, which endows them with a large surface/volume ratio and high specific area enabling to immobilize bigger amount of biomolecules onto biosensor surface [60].

Reusability of the antibody-based biosensors can be problematic because of the strength and irreversibility of antibody-antigen binding, and the regeneration of these surfaces without damaging the antibody layer can be complicated due to harsh conditions [61]. For impedance biosensors, extreme pH values of strong acids or bases may not be compatible with the chemistries employed for the protein immobilization, meaning that the reusability of a biosensor can be problematic. Radhakrishnan et al. studied the regeneration of antibody-based Si electrodes [62]. Even though they could regenerate the surfaces for 15 days, the impedance spectrum gradually degraded during these multiday regeneration trials.

Finally, it can be challenging to detect *small molecules* like heavy metals, pesticides, or antibiotic residues with EIS due to the exponential increase of the charge-transfer resistance through the polymer-protein layer to the underlying electrode surface [41, 63]. Small molecules (less than kDa) alone usually induce very small detectable response, which can be very difficult to measure especially in real samples where the concentration of the target molecule can be very low [41]. One possibility to improve the detection of small molecules is to conjugate these via a functional group to a larger carrier molecule (i.e., a protein) or with electrochemically bright metal and semiconductor nanomaterials, as changes due to binding of large molecules can be detected more easily detected. For example, Radhakrishnan et al. used impedance-based biosensor to detect two endocrine-disrupting chemicals (EDC) [41], which are small compounds found in various materials such as pesticides, additives, or contaminants in food [64]. It was found that for detecting small molecules, impedance biosensors can be operated at only one or a few frequencies that are most sensitive to analyte binding, and the sensitivity improved when attained with analyte conjugation. Gold nanoparticles (Au-NPs) have been used due to their electrochemically active surface; in particular, Au-NPs bound to the electrode digits disrupt the formation of the double layer around the electrodes, thus changing the double-layer capacitance [65–67]. For example, de Macedo et al. used Au-NPs for signal amplification, and comparing the results of free and conjugated protein, the latter generated a measured signal 40–50 times higher and the limit of detection 64 times lower [68]. MacKay et al. also used Au-NPs to evaluate the sensing ability of biosensor chips using impedance measurements and found that the adsorption of Au-NPs to the surface binding sites increased the impedance through double-layer capacitance and higher sensitivity is gained using single frequent measurement [65].

4. Applications of EIS-based biosensors in water analysis

A range of different EIS-based biosensing technologies for the detection of pollutants like pesticides and pathogens in water samples have been developed. A condensed overview of these biosensors including a brief description of the biosensor working principles, limit of detection, working range, and reproducibility is given in **Table 1**. Although not all these devices have been commercialized, they have been successfully tested in the laboratories.

4.1 Biosensors for pesticides and toxin analysis

Jiang et al. proposed an aptamer-based biosensor for the detection of acetamiprid [59]. To increase the sensitivity of the system silver nanoparticles, decorated nitrogen-doped graphene (NG) nanocomposites were used. This aptasensor exhibited a linear response in the range of 0.1 pM–1.0 nM and a detection limit of 0.01 pM. Zehani et al. developed two impedimetric biosensors for the detection of diazinon in aqueous medium using two different types of lipase, conjugated with BSA, immobilized onto functionalized gold electrodes [69]. Diazinon is one of the most commonly used organophosphate pesticides in the world, and lipase is used to specifically catalyze the hydrolysis of diazinon into diethyl phosphorothioic acid and 2-isopropyl-4-methyl-6-hydroxypyrimidine. The developed biosensors both presented linearity up to 50 μM with detection limit of 10 nM for *Candida rugosa*-based biosensor and 0.1 μM for porcine pancreas-based biosensor. They also studied the reproducibility and stability. Pichetsurnthorn et al. used nanoporous impedance-based biosensor for the detection of pesticide atrazine from river water [70]. To enhance the sensitivity of the system, nanoporous alumina was overlaid on the base surface of the metal electrode. The limit of detection for the detection of atrazine in river water and in drinking water was 10 fg/ml.

Zhang et al. constructed a three-dimensional (3D) graphene-based biosensor for microcystin-LR (MC-LR) detection and quantification in drinking water [54]. Microcystin-LR is a toxin produced by cyanobacteria. EIS was used for the electrochemical characterization of the biochemical action on the electrode-specific anti-MC-LR monoclonal antibodies for the selective detection of MC-LR. A detection limit of 0.05 mg/l was achieved, which is lower than that allowed limit proposed by the World Health Organization (WHO) (1 mg/l).

4.2 Biosensors for bacterial analysis

Mutreja et al. used impedimetric immunosensor for the detection of bacteria *Salmonella typhimurium* in water with detection limit 10^1 CFU/ml [71]. Graphene-graphene oxide screen-printed electrodes were functionalized with anti-OmpD antibodies to capture *Salmonella typhimurium* through its outer membrane protein OmpD. Barreiros dos Santos et al. presented an EIS-based biosensor for the detection of pathogen *Escherichia coli* O157:H7 in water [72]. The immunosensor detection limit was 2.0 CFU/ml, and linear working range was 10–10^4 CFU/ml. Rengaraj

Analyte	Sample	Recognition element	Electrode	LOD	Reproducibility	Response range	References
Acetamiprid	Wastewater	Aptamer with the following sequences: 5′-(SH)-(CH2)6-TGTAATTTGTCTGCAGCGGTTCTTGATCGCTGACACCATATTATGAAGA-3	Silver nanoparticles (NPs) decorated with nitrogen-doped graphene (NG) nanocomposites	33 pM	(RSD) 6.9% (n = 5)	10 pM–5 nM	[59]
Diazinon	River water	Lipase from *Candida rugosa* (CRL); lipase from porcine pancreas (PPL)	Functionalized gold electrode	10 nM (CRL); 0.1 μM (PPL)	(RSD) 2–5%	2–50 μM	[69]
Atrazine	River and bottled drinking water	Anti-atrazine antibodies	Nanoporous alumina membrane integrated with printed circuit board platform	10 fg/ml	—	10 fg/ml–1 ng/ml	[70]
Microcystin-LR (toxin produced by cyanobacteria)	Local tap water	Monoclonal microcystin antibodies (against ADDA, AD4G2, mouse IgG1)	3D-graphene-based biosensor (Ni/graphene composites coated with a PMMA solution)	0.05 μg/l	6.9% inter- and 3.6% intra-assay coefficients of variability	0.05–20 mg/l (R2 0.939)	[54]
Salmonella typhimurium species	Water	Anti-OmpD antibodies	Graphene-graphene oxide-modified screen-printed carbon electrodes	101 CFU/ml	—	—	[71]
Pathogen *Escherichia coli* O157:H7	Water	Anti-*E. coli* antibodies	Functionalized gold electrode	2 CFU/ml	(RSD) 2% (n = 3)	10–104 CFU/ml	[72]
Bacteria	Water	Lectin *concanavalin A*	Functionalized screen-printed electrode	103 CFU/ml		103–107 CFU/ml	[73]

Table 1.
The application, characteristics, and construction of impedance biosensors used in water analysis.

et al. fabricated an impedimetric paper-based biosensor for the detection of bacterial contamination in water [73]. They used lectin *concanavalin A* as a bioselective element due to its stability to interact with mono- and oligosaccharides on bacterial cells. The detection limit was approximately 1000 CFU/ml.

4.3 Biosensors for drug residue detection

A good overview about aptamer-based EIS biosensors to determine different groups of antibiotics in water samples is presented in Ref. [74].

Jacobs et al. use an EIS-based microdevice, coupled with a nanoporous membrane and functionalized antibodies, to detect erythromycin in different water sources—drinking water and river water [75]. The limit of detection in drinking water was found to be around 0.1 ppt. In milk the allowed maximum residue level for erythromycin is 40 ppb. In the river water, the sensitivity is usually lower because of the organic matter in it that can interfere with binding of erythromycin. The limit of detection in the river water samples was around 1 ppt. The overall impedance change was still large enough to show if the concentrations of erythromycin are in a range of suitable or unsuitable for drinking.

5. Conclusion

In this overview main challenges and limitations of impedance biosensors, including the complexity of impedance detection, susceptibility to non-specific binding, challenges with the sensitivity, limitations to small molecule, and reusability of the electrodes are analyzed.

Abbreviations

EIS	electrochemical impedance spectroscopy
IUPAC	International Union of Pure and Applied Chemistry
$I(t)$	current response
$V(t)$	applied voltage
(Δt)	time interval
φ	phase shift
$Z(\omega)$	impedance
$V(t)$	voltage-time function
$I(t)$	current-time function
DC	direct current
R_s	resistance of the electrolyte
CPE	constant phase element
R_{ct}	charge-transfer resistance
W	the Warburg impedance
C_{dl}	double-layer capacitance
BSA	bovine serum albumin
Au-NPs	gold nanoparticles
MC-LR	microcystin-LR
WHO	World Health Organization
NG	nitrogen-doped graphene
CFU	colony-forming unit

Author details

Kairi Kivirand[1,2]*, Mart Min[1] and Toonika Rinken[2]

1 Thomas Johann Seebeck Department of Electronics, Tallinn Technical University, Tallinn, Estonia

2 Institute of Chemistry, University of Tartu, Tartu, Estonia

*Address all correspondence to: kairi.kivirand@ttu.ee

References

[1] Goradel NH, Mirzaei H, Sahebkar A, Poursadeghiyan M, Masoudifa A, Malekshahi ZV, et al. Biosensors for the detection of environmental and urban pollutions. Journal of Cellular Biochemistry. 2018;**119**:207-212

[2] Sass JB, Colangelo A. European Union bans atrazine, while the United States negotiates continued use. International Journal of Occupational and Environmental Health. 2006;**12**:260-267

[3] Barth JAC, Steidle D, Kuntz D, Gocht T, Mouvet C, von Tümpling W, et al. Deposition, persistence and turnover of pollutants: First results from the EU project AquaTerra for selected river basins and aquifers. The Science of the Total Environment. 2007;**376**:40-50

[4] Kivirand K, Rinken T, Kagan M. Calibrating biosensors in flow-through set-ups: Studies with glucose Optrodes. In: Rinken T, editor. State of the Art in Biosensors. Rijeka: IntechOpen; 2013. pp. 331-351

[5] Nagel B, Dellweg H, Gierasch LM. Glossary for chemists of terms used in biotechnology. Pure and Applied Chemistry. 1992;**64**:143-168

[6] Syahir A, Usui K, Tomizaki K, Kajikawa K, Mihara H. Label and label-free detection techniques for protein microarrays. Microarrays. 2015;**4**:228-244

[7] Brosel-Oliu S, Uria N, Abramova N, Bratov A. Impedimetric sensors for bacteria detection. In: Rinken T, editor. Biosensors. Rijeka: IntechOpen; 2015. pp. 257-288

[8] Karube I, Nomura Y. Enzyme sensors for environmental analysis. Journal of Molecular Catalysis B: Enzymatic. 2000;**10**:177-181

[9] Nigam VK, Shukla P. Enzyme based biosensors for detection of environmental pollutants-a review. Journal of Microbiology and Biotechnology. 2015;**25**:1773-1781

[10] Tudorache M, Bala C. Biosensors based on screen-printing technology, and their applications in environmental and food analysis. Analytical and Bioanalytical Chemistry. 2007;**388**:565-578

[11] Wang X, Lu X, Chen J. Development of biosensor technologies for analysis of environmental contaminants. Trends in Environmental Analytical Chemistry. 2014;**2**:25-32

[12] Amine A, Mohammadi H, Bourais I, Palleschi G. Enzyme inhibition-based biosensors for food safety and environmental monitoring. Biosensors & Bioelectronics. 2006;**21**:1405-1423

[13] Mallat E, Barceló D, Barzen C, GauglitzG, AbukneshaR. Immunosensors for pesticide determination in natural waters. TrAC, Trends in Analytical Chemistry. 2001;**20**:124-132

[14] Ramírez NB, Salgado AM, Valdman B. The evolution and developments of immunosensors for health and environmental monitoring: Problems and perspectives. Brazilian Journal of Chemical Engineering. 2009;**26**:227-249

[15] Jiang X, Li D, Xu X, Ying Y, Li Y, Ye Z, et al. Immunosensors for detection of pesticide residues. Biosensors & Bioelectronics. 2008;**23**:1577-1587

[16] Felix FS, Angnes L. Electrochemical immunosensors – A powerful tool for analytical applications. Biosensors & Bioelectronics. 2018;**102**:470-478

[17] Justino CIL, Duarte AC, Rocha-Santos TAP. Recent progress in

biosensors for environmental monitoring: A review. Sensors. 2017;**17**:E2918

[18] Paniel N, Baudart J, Hayat A, Barthelmebs L. Aptasensor and genosensor methods for detection of microbes in real world samples. Methods. 2013;**64**:229-240

[19] Rapini R, Marrazza G. Electrochemical aptasensors for contaminants detection in food and environment: Recent advances. Bioelectrochemistry. 2017;**118**:47-61

[20] Mansfeld F. Electrochemical impedance spectroscopy (EIS) as a new tool for investigating methods of corrosion protection. Electrochimica Acta. 1990;**35**:1533-1544

[21] Jüttner K. Electrochemical impedance spectroscopy (EIS) of corrosion processes on inhomogeneous surfaces. Electrochimica Acta. 1990;**3510**:1501-1508

[22] Grundmeier G, Schmidt W, Stratmann M. Corrosion protection by organic coatings: Electrochemical mechanism and novel methods of investigation. Electrochimica Acta. 2000;**45**:2515-2533

[23] Majchrzycki W, Jankowska E, Baraniak M, Handzlik P, Samborski R. Electrochemicalimpedancespectroscopy and determination of the internal resistance as a way to estimate Lead-acid batteries condition. Batteries. 2018;**4**:2-13

[24] Wiart R. Elementary steps of electrodeposition analysed by means of impedance spectroscopy. Electrochimica Acta. 1990;**35**:1587-1593

[25] Sahlin, SL, Araya SS, Andreasen SJ, Kær SK. Electrochemical impedance spectroscopy (EIS) characterization of reformate-operated high temperature PEM fuel cell stack. International Journal of Power and Energy Research. 2017;**1**:20-40

[26] Yuan X, Wang H, Colin Sun J, Zhang J. AC impedance technique in PEM fuel cell diagnosis-A review. International Journal of Hydrogen Energy. 2007;**32**:4365-4380

[27] Park JS, Choi JH, Woo JJ, Moon SH. An electrical impedance spectroscopic (EIS) study on transport characteristics of ion-exchange membrane systems. Journal of Colloid and Interface Science. 2006;**300**:655-662

[28] ErtuğruL HD, Uygun ZO. Impedimetric biosensors for label-free and Enzymless detection. In: Rinken T, editor. State of the Art in Biosensors. Rijeka: IntechOpen; 2013. pp. 179-196

[29] Min M, Parve T, Pliquett UF. Impedance detection. In: Dongquing L, editor. Encyclopedia of Microfluidics and Nanofluidics. Boston: Springer; 2015. pp. 1338-1361

[30] Wang Y, Ye Z, Ying Y. New trends in impedimetric biosensors for the detection of foodborne pathogenic bacteria. Sensors. 2012;**12**:3449-3471

[31] Uygun ZO, Ertuğrul Uygun HD. A short footnote: Circuit design for faradaic impedimetric sensors and biosensors. Sensors and Actuators B: Chemical. 2014;**31**:448-453

[32] Daniels JS, Pourmand N. Label-free impedance biosensors: Opportunities and challenges. Electroanalysis. 2007;**19**:1239-1257

[33] Reddy B, Salm E, Bashir R. Electrical chips for biological point-of-care detection. Annual Review of Biomedical Engineering. 2016;**18**:329-355

[34] Prodromidis MI. Impedimetric immunosensors-a review. Electrochimica Acta. 2010;**55**:4227-4233

[35] Santos A, Davis JJ, Bueno PR. Fundamentals and applications of impedimetric and redox capacitive biosensors. Journal of Analytical and Bioanalytical Techniques. 2014;**S7**:016

[36] David N-H, Javier R-T, Min M. Fundamentals, recent advances and future challenges in bioimpedance devices for healthcare applications. Journal of Sensors. 2019;**5**:1-42

[37] Heo J, Hua SZ. An overview of recent strategies in pathogen sensing. Sensors. 2009;**9**:4483-4502

[38] Bogomolova A, Komarova E, Reber K, Gerasimov T, Yavuz O, Bhatt S, et al. Challenges of electrochemical impedance spectroscopy in protein biosensing. Analytical Chemistry. 2009;**81**:3944-3949

[39] Baryeh K, Takalkar S, Lund M, Liu G. Introduction to medical biosensors for point of care applications. In: Narayan RJ. Editor. Medical Biosensors for Point of Care (POC) Applications Woodhead Publishing. 2017. p. 3-25

[40] Morales MA, Halpern JM. Guide to selecting a biorecognition element for biosensors. Bioconjugate Chemistry. 2018;**29**:3231-3239

[41] Radhakrishnan R, Suni II, Bever CS, Hammock BD. Impedance biosensors: Applications to sustainability and remaining technical challenges. ACS Sustainable Chemistry & Engineering. 2014;**2**:1649-1655

[42] Frederix F, Bonroy K, Reekmans G, Laureyn W, Campitelli A, Abramov MA, et al. Reduced nonspecific adsorption on covalently immobilized protein surfaces using poly(ethylene oxide) containing blocking agents. Journal of Biochemical and Biophysical Methods. 2004;**58**:67-74

[43] Riquelme MV, Zhao H, Srinivasaraghavan V, Pruden A, Vikesland P, Agah M. Optimizing blocking of nonspecific bacterial attachment to impedimetric biosensors. Sensing and Bio-Sensing Research. 2016;**8**:47-54

[44] Contreras-Naranjo JE, Aguilar O. Suppressing non-specific binding of proteins onto electrode surfaces in the development of electrochemical immunosensors. Biosensors. 2019;**9**:15

[45] Chokchai P, Kamolpach R. Investigation of bovine serum albumin and casein adsorption on gold surface by using electrochemical impedance spectroscopy. Srinakharinwirot Science Journal. 2015;**31**:18-31

[46] Ortiz-Aguayo D, del Valle M. Label-free aptasensor for lysozyme detection using electrochemical impedance spectroscopy. Sensors. 2018;**18**:E354

[47] Xiao Y, Isaacs SN. Enzyme-linked immunosorbent assay (ELISA) and blocking with bovine serum albumin (BSA)-not all BSAs are alike. Journal of Immunological Methods. 2012;**384**:148-151

[48] Utrata-Wesoek A. Antifouling surfaces in medical application. Polimery. 2013;**58**:685-695

[49] Groß A, Richter M, Kubinski DJ, Visser JH, Moos R. The effect of the thickness of the sensitive layer on the performance of the accumulating NOx sensor. Sensors. 2012;**12**:12329-12346

[50] Comeau KD, Meli MV. Effect of alkanethiol chain length on gold nanoparticle monolayers at the air-water interface. Langmuir. 2012;**28**:377-381

[51] Campuzano S, Pedrero MM, Montemayor CC, Fatás E, Pingarrón JM, Fatas E, et al. Characterization of alkanethiol-self-assembled monolayers-modified gold electrodes by electrochemical impedance spectroscopy. Journal

of Electroanalytical Chemistry. 2006;**586**:112-121

[52] Bhadra P, Shajahan MS, Bhattacharya E, Chadha A. Studies on varying n-alkanethiol chain lengths on a gold coated surface and their effect on antibody–antigen binding efficiency. RSC Advances. 2015;**5**:80480-80487

[53] Braiek M, Rokbani KB, Chrouda A, Mrabet B, Bakhrouf A, Maaref A, et al. An electrochemical immunosensor for detection of Staphylococcus aureus bacteria based on immobilization of antibodies on self-assembled monolayers-functionalized gold electrode. Biosensors. 2012;**2**:417-426

[54] Zhang W, Han C, Jia B, Saint C, Nadagouda M, Falaras P, et al. A 3D graphene-based biosensor as an early microcystin-LR screening tool in sources of drinking water supply. Electrochimica Acta. 2017;**236**:319-327

[55] Mehta J, Bhardwaj N, Bhardwaj SK, Tuteja SK, Vinayak P, Paul AK, et al. Graphene quantum dot modified screen printed immunosensor for the determination of parathion. Analytical Biochemistry. 2017;**523**:1-9

[56] Bonanni A, Loo AH, Pumera M. Graphene for impedimetric biosensing. TrAC, Trends in Analytical Chemistry. 2012;**37**:12-21

[57] Aceta Y, del Valle M. Graphene electrode platform for impedimetric aptasensing. Electrochimica Acta. 2017;**229**:458-466

[58] Shao Y, Wang J, Wu H, Liu J, Aksay IA, Lin Y. Graphene based electrochemical sensors and biosensors: A review. Electroanalysis. 2010;**22**:1027-1036

[59] Jiang D, Du X, Liu Q, Zhou L, Dai L, Qian J, et al. Silver nanoparticles anchored on nitrogen-doped graphene as a novel electrochemical biosensing platform with enhanced sensitivity for aptamer-based pesticide assay. The Analyst. 2015;**140**:6404-6411

[60] Tîlmaciu C-M, Morris MC. Carbon nanotube biosensors. Frontiers in Chemistry. 2015;**27**:1-21

[61] Goode JA, Rushworth JVH, Millner PA. Biosensor regeneration: A review of common techniques and outcomes. Langmuir. 2015;**31**:6267-6276

[62] Radhakrishnan R, Suni II. Antibody regeneration on degenerate Si electrodes for calibration and reuse of impedance biosensors. Sensing and Bio-Sensing Research. 2016;**7**:20-24

[63] Bradbury CR, Zhao J, Fermín DJ. Distance-independent charge-transfer resistance at gold electrodes modified by thiol monolayers and metal nanoparticles. Journal of Physical Chemistry C. 2008;**112**:10153-10160

[64] Bergman Å, Heindel J, Jobling S, Kidd K, Zoeller RT. State-of-the-science of endocrine disrupting chemicals, 2012. Toxicology Letters. 2012;**S17**:53

[65] MacKay S, Abdelrasoul G, Tamura M, Lin D, Yan Z, Chen J. Using impedance measurements to characterize surface modified with gold nanoparticles. Sensors. 2017;**17**:2141

[66] Wang J, Profitt JA, Pugia MJ, Suni II. Au nanoparticle conjugation for impedance and capacitance signal amplification in biosensors. Analytical Chemistry. 2006;**78**:1768-1773

[67] Li Y, Schluesener HJ, Xu S. Gold nanoparticle-based biosensors. Gold Bulletin. 2010;**43**:29-41

[68] de Macedo EF, Formaggio DMD, Santos NS, Tada DB. Gold nanoparticles used as protein scavengers enhance surface plasmon resonance signal. Sensors. 2017;**17**:2765

[69] Zehani N, Dzyadevych SV, Kherrat R, Jaffrezic-Renault NJ. Sensitive impedimetric biosensor for direct detection of diazinon based on lipases. Frontiers in Chemistry. 2014;**2**:44

[70] Pichetsurnthorn P, Vattipalli K, Prasad S. Nanoporous impedemetric biosensor for detection of trace atrazine from water samples. Biosensors & Bioelectronics. 2012;**32**:155-162

[71] Mutreja R, Jariyal M, Pathania P, Sharma A, Sahoo DK, Suri CR. Novel surface antigen based impedimetric immunosensor for detection of salmonella typhimurium in water and juice samples. Biosensors & Bioelectronics. 2016;**85**:707-713

[72] Barreiros dos Santos M, Agusil JP, Prieto-Simón B, Sporer C, Teixeira V, Samitier J. Highly sensitive detection of pathogen Escherichia coli O157: H7 by electrochemical impedance spectroscopy. Biosensors & Bioelectronics. 2013;**45**:174-180

[73] Rengaraj S, Cruz-Izquierdo A, Scott JL, Di Lorenzo M. Impedimetric paper-based biosensor for the detection of bacterial contamination in water. Sensors and Actuators B: Chemical. 2018;**265**:50-58

[74] Mehlhorn A, Rahimi P, Joseph Y. Aptamer-based biosensors for antibiotic detection: A review. Biosensors. 2018;**8**:54

[75] Jacobs M, Nagaraj VJ, Mertz T, Selvam AP, Ngo T, Prasad S. An electrochemical sensor for the detection of antibiotic contaminants in water. Analytical Methods. 2013;**5**(17):4325-4329

Chapter 6

Dynamic Luminescent Biosensors Based on Peptides for Oxygen Determination

Geovany Albino de Souza, Tatiana Duque Martins and Flavio Colmati

Abstract

This chapter summarizes the fundamentals of biosensing techniques based on fluorescence spectroscopy and the protagonism of state-of-the-art luminescent biosensors in a wide range of scientific areas, from environmental monitoring to diagnostics and decease treatment, focusing on the paramount contribution of biosensors based on the Förster Resonance Energy Transfer (FRET) transducing mechanism. State-of-the-art FRET biosensors are specially characterized by outstanding sensitivity toward a number of environmental pollutants and dissolved oxygen in aquatic ecosystems, capable of detecting concentrations in the nano and picomolar scales. These biosensors have also been showing impressive performance over other methods in the study of real-time biological processes *in vivo* relevant to help understanding decease progression like cancer.

Keywords: fluorescent biosensors, ROS detection, sensitivity, FRET biosensors, new materials

1. Introduction

Fluorescence spectroscopy has been revolutionizing the field of life sciences and clinical routines such as diagnostics and biosensing, due to its impressive sensitivity and the biocompatibility of many fluorescent organic compounds, which allows one to probe biological processes in vivo in noninvasive bioimaging procedures. The improvement of instrumentation has granted optical-based sensing routines a new level of sensitivity, accuracy, and reliability. Very subtle changes in fluorescence intensity, or even extremely low levels of light that might result from an interaction between the fluorescent probe or sensor with the environment under study, are easily detectable nowadays thanks to modern instrumentation such as photomultiplier tubes, in which electronic impulses are created by just a single photon. In the time-resolved fluorescence approach, in which the fluorescence lifetime of the fluorophore is monitored, with current femtosecond pulsed lasers, even more sensitive and reliable measurements are possible, since the intensity decay is not affected by a number of possible undesired factors that interferes in the steady-state intensity. These advantages make the fluorescence spectroscopy a powerful technique, paving the way to the most wished rapid and low-cost sensing in a wide range of biological and environmental applications and point-of-care diagnostics for real-time

 IntechOpen

monitoring of physiological conditions. Sensing methods of remarkable sensitivity, reliability, and selectivity based on fluorescence spectroscopy dominates the field of sensing and biosensing. DNA sequencing and fragment analysis, fluorescence staining for bioimaging and fluorescence immunoassays are all based on fluorescence techniques.

The countless possibilities of combinations of biorecognition element, support matrix, and the transducing method in biosensors constituted of nanomaterials make it possible to design versatile and selective biosensors. In this review, particular attention is centered on luminescent biosensors based on the Förster resonance energy transfer, or FRET, biosensing transducing method, which encompasses a huge variety of biosensors, due to their unique sensitivity, selectivity, and fast response. For this reason, FRET-based sensors have enabled, for example, intracellular monitoring of ROS kinetics and oxygen sensing, which is vital for elucidating how tumor cells respond to treatment, in order to develop better therapeutic strategies. Before FRET sensors were introduced, these practices were hampered by difficulties and unreliability in real-time monitoring intracellular ROS. In fluorescence bioimaging, thanks to its high spatial and temporal resolution, it is becoming possible to probe in real time biological elements and processes, such as enzymatic reactions, protein/protein, protein/nucleic acid, protein/substrate, and biomembrane interactions.

In the context of environmental monitoring, with the possibility of miniaturization of biosensors based on nanomaterials, it is becoming possible to perform fast and accurate field analysis and real-time surveillance of analytes relevant in the assessment of water quality. A variety of FRET-based biosensors for water pollutants, such as heavy metal ions, pesticides, antibiotics, and halogenated compounds, is reported, some of them capable of detecting concentrations in the pico and nanomolar scales. Such sensitivity levels are far from reach with conventional analytical methods.

2. Principles and evolution of biosensing techniques

Biosensors are devices that can perform the measurement of a physiological activity in living organisms or that are constructed upon biological components. They determine chemical or biological analytes in systems where the minimum human intervention is present. They generate optical and/or electrochemical signals that are transduced by a variety of transducers, and depending on their operation, biosensors are classified.

To fulfill an application, biosensors can be constructed by using a wide variety of bioreceptors, that can deliver distinct types of signals and the choice of transducers and interfaces will respond for their selectivity and sensibility, as well as their configuration versatility and possibility of miniaturization. Techniques such as voltammetry, amperometry, potentiometry, among others, are exploited to transduce electrochemical signals, whereas fluorescence, light absorption or light reflectance in the ultraviolet (UV), visible, or near-infrared (NIR) spectral regions can have their intensity or lifetime changes determined to efficiently detect optical responses in optical biosensors. Also, there is a variety of biosensors that can explore dual or multiple transducers to deliver electrical and luminous signals that can be interpreted together or separately, giving rise for more versatile and usable biosensors.

Independent of their usage and characteristics, the basic configuration of the biosensors is similar. They must be composed of a sample holder, which is adapted to the sample physical characteristics; a biological recognition element, which must be highly selective; and a physical transducer to generate a measurable signal, a signal processor, and an interface that is able to communicate the data to the operator. The types of biosensor that can be constructed are based on the recognition elements employed, which can be any biological system, from antibodies to microbes and cells. It is selected considering the information to be obtained and the physical characteristics of the biosensor, which in its turn, determines the durability and the more applicable processes to construct the biosensor.

The specificity and the sensitivity of biosensors have been the concern of several scientists due to the variety of biochemical processes that need to be evaluated and followed, the increase of accuracy and reproducibility of measurements that is fundamental for the spread of usage of such devices, and due to the need for miniaturizing and automatizing devices, in order to turn them more applicable to distinct regions of a living organism.

In special, optical biosensors can be widely applied if they accompanied the development of the spectroscopic and microscopic technique development to improve their transducing methods, signal processing, and interface. If these components are well developed, optical biosensors and, in special, fluorescent biosensors can present high sensitivity, perform real-time measurements with high frequency of detection, which enable them to find application in diagnostics and therapeutics, with the right transducers, it is possible to image disease progression and to monitor the organism response to therapeutics, and also, they can be thought for drug discovery programs development, as well as for clinical evaluation of new drugs [1]. In recent years, recognition elements based on graphene had been widely used due to their excellent electrical and optical properties. In biosensing, materials based on graphene had promoted efficient detection of biomarkers and have proportioned an important technological advance, due to the perspective of developing new and interesting materials, such as graphene-like 2D materials and the impressive single-atom-thick layers of van der Waals materials [2].

Most of the fluorescent biosensors are small molecules that are arranged on a receptor that identifies a specific target, and its fluorescent response is readily transduced. In this case, the signal recognition is based on the distinct fluorescence emitted by the biosensor in the presence of the analyte, which can be metabolites, proteins, ions, or antibodies. These biosensors are based on the steady-state fluorescence that the device can produce. They find effective application for early biomarker detection, for instance, in clinical diagnostics and ordinary biochemical processes, as well as for tissue imaging and, as an extrapolation, in image guided surgery.

Biosensors based on time-resolved fluorescence are promising because they are able to promote improvements on selectivity, specificity, and sensibility, becoming ultrasensitives, capable of determining minimal local variations of the analyte concentration, and can be combined to several other analytical techniques. In this context, there are the nanopores, which are highly sensitive biosensors, able to detect the analyte at the range of nanomolar of concentration, due to their characteristic structure of a nanometer scale pore that is similar to the size of the target molecule, enabling a single-molecule analysis [3].

Any of the fluorescence properties of the recognition element, namely, the intensity, wavelength, anisotropy or lifetime, can be exploited in optical sensing. One straightforward mechanism to consider is collisional quenching, in which a fluorophore has its fluorescence quenched upon collision with the analyte molecule. Nevertheless, the most encountered and relevant mechanism in sensing is the Förster Resonance Energy Transfer (FRET), which occurs via long range dipole-dipole interaction when an energy donor, in an excited state, and an acceptor are brought into close proximity, but not necessarily into contact. In FRET, the

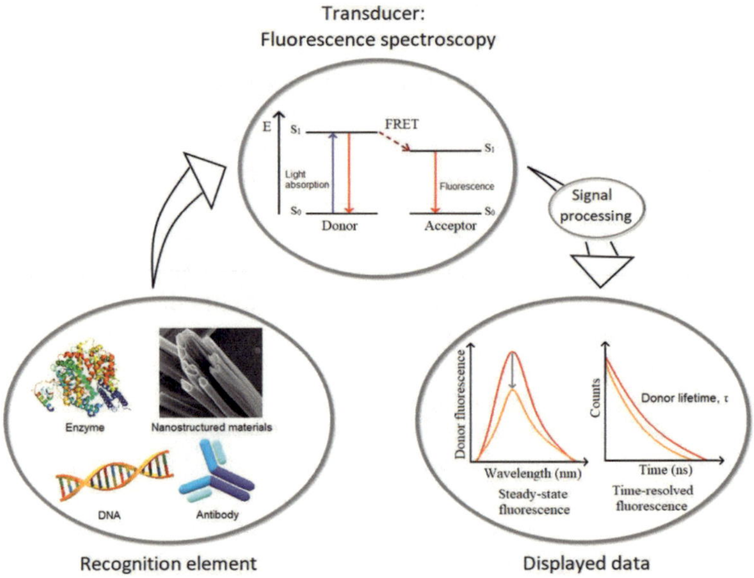

Figure 1.
Illustration of the components and function of biosensors based on fluorescence spectroscopy transducing.

fluorescence intensity and decay time of the donor are decreased, as depicted in the Jablonski diagram and spectrum models in **Figure 1**. The energy transfer efficiency in FRET depends on the distance between donor and acceptor: when the donor-acceptor pair is in between 20 and 60 angstroms apart, which is called the Förster radius, the efficiency of energy transfer is around 50%. The efficiency also depends on the spectral overlap between the absorption spectrum of the acceptor and the emission spectrum of the donor: the greater the overlap, the more efficient the process.

Since FRET does not require contact between the electronic clouds of donor and acceptor, it can occur over macromolecular distances. This is one of the reasons, along with energy transfer efficiency, responsible for the high sensitivity of FRET-based sensors. Low concentrations of analyte that would result in great distances from the donor-acceptor pair would most likely involve FRET rather than collisional quenching, which requires physical contact, in order to bring about a change in the FRET process necessary for sensing. Besides, since the donor and acceptor in a FRET sensor do not need to be bound molecules, it simplifies the design of the fluorophore because the donor is not required to be intrinsically sensitive to the analyte and can be chosen according to the desired light source [4].

A great variety of FRET sensors and biosensors can be found in the literature. The major advantages that make them special are high sensitivity and fast response. Besides, due to the biocompatibility of FRET-based biosensors, allied to the inherent sensitivity of optical sensing techniques, they are becoming ubiquitous in clinical applications and in the field of biomedical research. A recent example is a fluorescent peptide/GO sensor containing a fluorophore-labeled peptide sequence that proved versatile for measuring the activity of different protein kinases. Kinases are group of proteins that regulates intracellular phosphorylation pathways. Several deceases such as cancer, diabetes, Alzheimer's, etc. are related to anomalous activity of kinase proteins. In their sensor, the fluorophore-labeled peptide sequence is cleaved by the carboxypeptidase, in the absence of phosphorylation, and is separated from the GO, resulting in recovery of fluorescence [5].

Quantum dots (QDs) comprise another class of intensively researched materials for biosensing application due to their advantages such as high photostability and large extinction coefficient. Additionally, their broad absorption spectra and the possibility of tuning emission wavelength make them suitable as sensors based on FRET energy transfer, in which the spectral overlap is important for efficiency of the process, and, ultimately, the sensitivity of the biosensor. Their broad absorption spectrum also enables selective excitation of QD donors without exciting the acceptors and allows excitation of different donors at once, making them useful for multiplexing applications. Their high luminescence and nanoscale dimensions also make them ideal for bioimaging applications as fluorescent probes [6].

Despite their versatility and large scope of possibilities for biosensing, one major disadvantage of QDs fabricated with inorganic materials is their toxicity, which limits their clinical applications. Graphene and other carbon-based QDs, unlike the conventional heavy-metals DQs, are biocompatible, environment friendly, and easier to prepare [7, 8], and for that reason, they have gained considerable attention. Among the most recent examples of CQD-based is a fluorescent CQD biosensor for hydrogen peroxide detection and simultaneous monitoring of the acetylcholinesterase reaction system [7]. A great number of QD-based biosensors for environmental monitoring are found in the literature. A few recent contributions are described in Section 4.

Similarly to graphene QDs, the high efficiency of energy transfer from dyes to graphene oxide, GO, along with GO's intrinsic properties, opened up a new avenue for designing a lot of FRET-based biosensors. Thanks to pi stacking and hydrogen bonding interactions, GO is capable of strong binding with biomolecules, such as fluorescent dyes, which are quenched by GO via the FRET process. In the past few years, a number of GO-based biosensors using DNA as a probe are reported [9, 10].

As a special case of FRET-based biosensors, there is the bioluminescence resonance energy transfer (BRET) principle, which has been used to produce new and ultrasensitive biosensors. In these biosensors, a bioluminescent enzyme is the energy donor and a specific fluorescent molecule, chosen by spectral overlapping, acts as an acceptor. This process is extensively used to monitor and study molecular interactions between proteins and other metabolites, in vitro and in living cells [11].

3. ROS and oxygen sensing

One very important class of compounds that play a major role in regulating biological processes, which also have a close relationship with the differentiated metabolism profile of tumor cells, is the reactive oxygen species, ROS, and for that reason, they receive significant attention in sensing/biosensing research. ROS are very reactive free radicals that act as electron acceptors, thus being strong oxidizing agents, which react with any neighboring molecule in order to attain a stable configuration. Hydrogen peroxide, the superoxide anion, and the singlet excited states of oxygen are examples of ROS. These molecules are produced physiologically mainly as a by-product of oxygen metabolism during electron transfer events in respiratory chain processes. Since they are highly reactive, ROS are harmful for the cells, and antioxidant enzymes located in the cytosol and mitochondria are responsible for a delicate regulation process that control the oxidative stress generated by ROS. Despite their toxic effects, moderate levels of ROS play a role in vital biological processes, such as biological signaling, chemical defense, biosynthetic reactions, etc. [12]. For instance, in biological signaling, the ROS act as secondary messengers in cellular adhesion, spreading, and migration, thus governing the proliferation and, ultimately, the survival of cells [13].

Multicellular organisms also maintain a homeostasis of O_2 levels by regulating the distribution of O_2 according to the demands of each organ. In this context, the cells also respond to hypoxia condition, a state of insufficient levels of O_2 necessary for maintaining nominal cellular function, in an effort to adapt to such condition and ensure its survival. This condition is always present in tumor cells, and understanding the cell mechanisms of coping with hypoxia is crucial for understanding tumor growth and survival. For that to work, the cells have a sensing mechanism for O_2. The hypoxia-inducible factors HIF-1 and HIF-2 are oxygen sensitive transcriptional complexes that mediate the response of cells to O_2 levels [14].

It has been reported that the level of ROS in the cells is related to the activity of HIF-α factor. However, elucidation of cell metabolism related to or governed by ROS, as well as the cellular oxygen sensing mechanisms, has been hampered by the difficulties in tracking intracellular ROS kinetics with reliability using fluorescent dyes. FRET-based sensors have demonstrated to overcome these obstacles [14–16]. Guzy et al. suggested that mitochondria functions as an O_2 sensor and signal hypoxia-induced HIF stabilization by releasing ROS to the cytosol. To confirm this debated hypothesis, they fabricated a reliable and accurate redox-sensitive FRET protein sensor for intracellular ROS determination that consists of a cyan and yellow fluorescent protein as the donor-acceptor pair linked by a redox-regulated HPS33 protein domain. Oxidation of the protein domain by ROS causes a conformational change, which alters the donor-acceptor distance required for FRET efficiency, thus leading to fluorescence recovery [14]. More recently, Bernardini et al. also used this same sensor configuration for monitoring the kinetics of ROS in cultured cells of mouse carotid body, seeking to elucidate the role of a NADPH oxidase subunit on cell response to O_2 levels [15].

Overall, cyan and yellow fluorescent protein FRET sensors containing an internal metabolite-binding protein are successfully used for monitoring biomolecules in real time, metabolite dynamics, protein interaction, and signal transduction, due to the fast response ability of these sensors. Just like in aptamer sensors, the choice of the right metabolite-binding protein is the key for obtaining a highly selective FRET sensor for the desired target [16].

Apart from reactive oxygen species, the O_2 itself is also an important indicator of biological processes, including the monitoring of decease progression. A variety of optical biosensors for oxygen determination are found in the literature, given the importance of this molecule in the study of biological systems.

Recently, fluorescence biosensors based on QDs are proving their potential for monitoring tumor activity, thanks to advantages such as the ability to penetrate solid tumor cells, high photoluminescence quantum yields, photostability, and the other photophysical properties already mentioned. Oxygen-sensitive phosphorescent molecules are particularly interesting for detecting oxygen in biological systems, because it is noninvasive and has high spatiotemporal resolution [17].

The underlying energy transfer mechanism involved in oxygen sensing is mostly a triplet-triplet annihilation process. After absorption of light, the sensing molecule is excited to a singlet excited state, S_1. Subsequently, it can return to the ground state, S_0, through fluorescence or undergo intersystem crossing to a triplet excited state, T_1, and then return to the S_0 state by phosphorescence. Alternatively, in a competing process, the molecule in the T_1 state can interact with molecular oxygen in the ground triplet state via collisional quenching. When this happens, a triplet-triplet annihilation process occurs, which is characterized by generation of excited-state singlet oxygen, as illustrated in the Jablonsky diagram in **Figure 2**. This process involving the O_2 is also called a photosensitization effect.

Consequently, a quenching of fluorescence or phosphorescence (depending on what radiative process is being monitored) is observed. The lifetime is

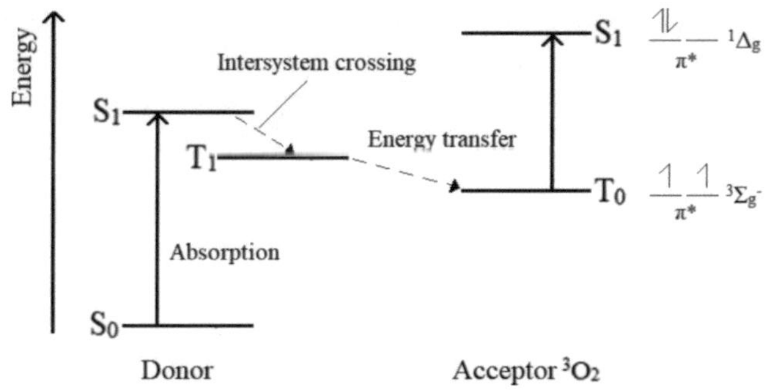

Figure 2.
Jablonsky energy diagram illustrating a T-T annihilation process involving a photosensitizer molecule and molecular oxygen.

also decreased. The oxygen concentration can be determined in a steady-state or time-resolved manner, in which the intensity or lifetime decrease of the quenched molecule is related to the concentration of oxygen by the Stern-Volmer kinetics of collisional quenching, as shown below.

$$\frac{I_0}{I} \text{or} \frac{\tau_0}{\tau} = 1 + kq\,\tau_0[Q](1) \quad (1)$$

In this expression, I_0 and I are the fluorescence intensities in the absence and presence of O_2, respectively, τ_0 is the natural fluorescence lifetime of the molecule in the absence of O_2, τ is the lifetime at a given oxygen concentration, [Q], and kq is the bimolecular quenching rate constant [4].

However, intersystem crossing to T_1 state is generally a slow process, and in order to achieve an oxygen sensor of high sensitivity, the population of the triplet state must be maximized. The addition of heavy metal atoms can circumvent this limitation by increasing spin-orbit coupling, which favors the forbidden transition between electronic states of different spin multiplicities. Still, once in the T_1 state, the phosphorescence competes with energy transfer to O_2, which further limits the sensitivity of the sensor. The coupling of an oxygen-responsive phosphorescent molecule with a QD results in a sensor with enhanced sensitivity toward oxygen. FRET readily occurs between the molecule and QD, which form the donor-acceptor pair, and emission becomes oxygen-dependent due to the analyte interference in the FRET process.

In addition to FRET, QD sensors can also function through Dexter exchange interactions, another mechanism of nonradiative energy transfer between an acceptor-donor pair, which involves the donation of an electron from the LUMO orbital of the donor followed by the transfer of an electron from the acceptor HOMO orbital to the HOMO of the donor. Unlike the FRET, Dexter interactions require physical contact between donor and acceptor. Both mechanisms can compete depending on the degree of spectral overlap. It has been found that the size of QD determines the predominant energy transfer mechanism in QD-based sensors: smaller QD, which exhibit low spectral overlap and high orbital overlap with the donor, favors Dexter energy transfer, whereas in larger QDs, FRET dominates [17].

Belonging to the realm of QD-based sensors for oxygen, self-assembled sensors have interesting advantages including ease of preparation and allow easy fine-tuning of donor-acceptor ratio. In general, these sensors are comprised of a metal ion bound to functional groups such as imidazole, amines, etc. and an acceptor molecule attached to the functional group. For instance, Lemon et al. managed to red shift the emission of a CdSe QD sensor reported previously in order to increase its sensitivity at physiological O_2 pressure by pairing Pd(II) porphyrins, which emits at 690 nm, with the CdSe core-shell QDs that emit at 519 nm. The QD was chosen to maximize spectral overlap with Pd(II) porphyrin absorption, thereby increasing FRET efficiency to 94%, greatly improving sensitivity [18]. Later on, other authors have used Au(III) corroles to shift the emission even further to the NIR [19]. Red emitting sensors are interesting for biomedical applications due to the greater penetration of red light into organic tissues and less scattering. These contributions reveal the versatility of QD-based sensors, which can be easily designed and adjusted to fulfill the desired purpose.

Fluorescence microscopy is another common technique in biological and clinical fields for visualizing intracellular structures, both in vitro and in vivo, which is based on the staining of a cell with a fluorescent probe. It is also possible to determine intracellular concentration of analytes of interest and monitor reactions. However, one major problem of microscopy based on steady-state intensity is the intensity dependence on the probe concentration. The difficulty in knowing the probe concentration within the different regions of the cell impedes quantitative measurements with reliability. Fluorescence lifetime imaging, on the other hand, circumvents this problem because the lifetime of the fluorophore probe is independent on its concentration. Therefore, variations in lifetime due to interactions of the probe with biomolecules can be correlated to analyte concentration regardless of the probe concentration. For this reason, high fidelity images with improved contrasts can be achieved. Lifetime imaging is employed, for example, in intracellular oxygen sensing, which is not possible via any microscopic method based on intensity measurements [4].

4. FRET applied to environmental biosensing

Due to the outstanding selectivity and sensitivity of optical biosensors, especially those based on FRET transducing, they meet the requirement of trace and even ultratrace detection of a great variety of environment pollutants, such as pesticides, antibiotics, halogenated contaminants [20], etc., which represent a major concern of the modern era due to the threat they pose to ecosystems and human health. In addition to this, the portability offered by the possibility of miniaturization of biosensor platforms enables the fast and low-cost field analysis, which is not possible through expensive conventional analytical methods like chromatography, mass spectrometry, and others.

Aptamer FRET sensors, or aptasensors, for instance, comprise a class of versatile and very sensitive biosensors, capable of detecting concentrations in nano and picoscales. Ultrasensitive FRET aptasensors for trace detection of metal ions [21, 22] and antibiotics [23] are reported. In a multiplexed detection system for Pb(II), Hg(2), and Ag(II) ions, the binding of an ion or multiple ions to DNA sequences triggers the DNA self-assembly. Subsequently, a cascade FRET event results in a fluorescent spectrum that can be interpreted as a fingerprint of the presence of a single or multiple metal ions [21]. In the sensor for the kanamycin antibiotic, an aptamer sensor, upconversion nanoparticles bound to an aptamer for kanamycin act as energy donor and the graphene as acceptor, in which the FRET is blocked in the presence of kanamycin, resulting in fluorescence. An impressive lower detection limit of 9.0 picomolar concentration is reported in aqueous buffer solution [23]. Indeed, by designing the suitable aptamer, versatile and selective FRET sensors for countless targets can be constructed.

Quantum dot and nanoparticle biosensors of equally impressive sensitivity for molecules of a wide range of sizes, from ions to large proteins, temperature, pH, and oxygen, also have potential application in environmental monitoring [24–26]. Most recent examples include a sensor for edifenphos fungicide, comprised of a ZnS QD conjugated to a single-stranded DNA aptamer immobilized on a graphene oxide sheet. In this sensor, FRET occurs between the QD and graphene sheet, and the fluorescence quenching is proportional to analyte concentration. Interestingly, their sensor showed remarkable selectivity, even when comparing to other pesticides of similar molecular structure [27]. In a chlorophyll-containing carbon QD of tunable fluorescence for organophosphate pesticide determination, the fluorescence of the QD is quenched via a FRET process by gold nanoparticles when the analyte is not present [28]. In another recent contribution, Luo et al. constructed a highly sensitive fluorescent sensor of Au/Ag nanoparticles containing rhodamine B for the detection of organophosphorus pesticides, which showed a detection limit of 1.8-pg ml^{-1} in fruit and water samples [25]. A similar biosensor of Au/Ag core-shell nanoparticles for the detection of arsenic has a lower detection limit of 0.1 ppb (parts per billion) [29]. Another FRET sensor constituted of Au upconversion nanoparticles for fluoroquinolones detection showed a sensitivity of 0.19–0.32 ng ml^{-1}. Fluoroquinolones are a class of antibiotics that have become serious water contaminants [30].

Other prominent biosensors for environmental surveillance are the nanophotonic biosensors, which are devices constituted of biological receptor layers immobilized onto the core surface of a waveguide, to detect evanescent waves [31]. Their functioning mechanism is based on the exposure of the waveguide surface to the analyte, resulting in a biochemical interaction that promotes a local change in the optical properties of the waveguide transducer, which can be detected, and its amplitude is modulated by the concentration of the analyte. An advantage of photonic biosensors is that they can be integrated to lab-on-a-chip platforms, enhancing their application possibilities.

Oxygen sensing is not limited to the biomedical field. It is also a valuable analyte in environmental monitoring. Determination of O_2 levels in aqueous ecosystems, such as rivers and lakes, is a common routine for evaluating the habitability conditions of these waters. Biochemical oxygen demand (BOD) and chemical oxygen demand (COD) analyses are the standard quantitative analytical methods applied for that purpose. BOD is the amount of dissolved oxygen demanded by aerobic microorganisms to decompose organic matter in a given water sample during an incubation time, usually 5 days. BOD expresses the concentration of consumed oxygen during this time period. In rivers polluted with high levels of organic waste, aerobic bacteria consume the dissolved O_2 during decomposition of organic matter, which results in a drastic reduction of available O_2 that aquatic animals need to survive. Analogous to BOD, the COD is a more general method, which gives the amount of oxygen needed for oxidation of any chemically oxidizable material, apart from organic matter [32].

Biosensors for oxygen and organic pollutants sensing for environmental surveillance are evolving rapidly due to many advantages they offer over the traditional methods of BOD and COD analyses, such as faster and more accurate results and the possibility of online and real-time monitoring of water quality [32–34]. One recent contribution in this field includes a microbial fuel cell biosensor for real-time BOD analysis that was tested for urine sensing. The device emits a sound alarm whenever the concentration of the analyte exceeds a given concentration threshold and is self-powered by the electroactive microorganisms of the microbial fuel cell [34].

Regarding oxygen sensing, both electrochemical and optical biosensors of noticeable sensitivity are found in the literature. In a recent electrochemical biosensor, peptide micro/nanostructures are self-assembled with a complex of copper that acts as the oxygen reduction catalyst, immobilized onto a glassy carbon. This biosensor showed a lower detection limit of 0.1 mg l^{-1} [35]. Most recent optical biosensors based on FRET transducing include a BOD biosensor chip and a ratiometric FRET sensor. In the biosensor chip for BOD analysis, an oxygen sensitive ruthenium complex coated with a polyethylene-polypropylene film permeable only by oxygen avoids the interference of pollutants from the sample. In this biosensor, the fluorescence intensity is correlated with oxygen concentration [36]. Another ratiometric FRET oxygen sensor consists of a Pt(II)-5,10,15,20-tetrakis-(2,3,4,5,6-pentafluorophenyl)-porphyrin oxygen probe entrapped in a copolymer matrix that is capable of real-time monitoring of extra-cellular O_2 consumption by *E. coli* bacteria and Hela cells. This biosensor showed a sensitivity of 0.08 mg l^{-1} [37]. The coating of the sensing unit or its immobilization in a matrix selective to oxygen permeability is a commonly adopted strategy in the design of optical sensors for oxygen in order to ensure its selectivity. Additionally, transition metal complexes, especially those of ruthenium and platinum, have a long phosphorescence lifetime, a requirement for efficient energy transfer from the sensing unit to molecular oxygen through collisional quenching, as described in Section 3, necessary for achieving high levels of sensitivity [4, 38, 39].

Our group has also developed a colorimetric sensor for dissolved O_2. Our sensor, comprised of a self-assembled peptide containing a fluorescent dye, is based on a FRET energy transfer between the constituents of the system that arises from the formation of a charge transfer complex. It showed remarkable sensitivity and selectivity toward dissolved O_2, both in steady-state and time-resolved fluorescence measurements. This self-assembled sensing platform, which was tested in fish breeding environment and showed good reproducibility, might be useful in analytical methods for determination of O_2 levels in polluted water samples.

Additionally, our material, when allied with an antioxidant drug used in cancer treatment, showed antioxidant activity by sensing singlet oxygen, as well as pro-oxidant behavior by generating that same reactive oxygen species when irradiated with light, which makes it promising for photodynamic therapy as well [40]. The singlet excited state of O_2 is perhaps the most important of the ROS molecules. Due to its considerable lethal effect for cells, it is exploited in photodynamic therapy, an alternative approach for a number of cancers that has proven to be efficient and far less invasive and harmful than the side effects of conventional treatment protocols. It is based on the same photosensitization process as described earlier.

5. Conclusions

The state-of-the-art contributions summarized in this chapter reveal the undoubtful vanguard of luminescent biosensors in the race toward new low-cost, biocompatible, and smart materials to help solving some of the most relevant problems the modern civilizations face, such as environmental pollution and diseases like cancer.

Biosensors based on optical techniques, allied with biological molecules and nanomaterials, such as quantum dots, are constantly bringing a new family of versatile sensors and biosensors that are providing unprecedented levels of accuracy, sensitivity, and control in the study of biological processes relevant in disease treatments and point-of-care devices for environmental monitoring. Regarding versatility, the design of aptamer sensors and quantum dot conjugate systems, for instance, allows countless modifications and combinations that can be easily carried out in order to fulfill the specificity of the desired purpose. The current pace in the development of this new generation of versatile, adaptive biosensors based on nanoscale and biological materials certainly promises to solve the main issues in biosensing development, which is the specificity, sensitivity, and portability requirements for the spread use of biosensing devices.

Acknowledgements

Authors thank FAPEG for scholarship and CNPq for financial support.

Conflict of interest

The authors acknowledge that there is no conflict of interests in this work.

Author details

Geovany Albino de Souza, Tatiana Duque Martins* and Flavio Colmati
Chemistry Institute, Federal University of Goias, Goiânia, Brazil

*Address all correspondence to: tatiana@ufg.br

References

[1] Mehrotra P. Biosensors and their applications—A review. Journal of Oral Biology and Craniofacial Research. 2016;**6**:153-159. DOI: 10.1016/j.jobcr.2015.12.002

[2] Bollella P, Fusco G, Tortolini C, Sanzã G, Favero G, Gorton L, et al. Beyond graphene: Electrochemical sensors and biosensors for biomarkers detection. Biosensors and Bioelectronics. 2017;**89**(part 1):152-166. DOI: 10.1016/j.bios.2016.03.068

[3] Lee K, Kyeong-Beom P, Hyung-Jun K, Jae-Seok Y, Chae H, Hyun-Mi K, et al. Recent progress in solid-state nanopores. Advanced Materials. 2018;**30**:28. Artical ID: 1704680. DOI: 10.1002/adma.201704680

[4] Lakowicz JR. Principles of Fluorescence Spectroscopy. 3rd ed. New York: Springer; 2006. 954 p. ISBN 0-387-31278-1

[5] Zhou J, Xu X, Liu W, Liu X, Nie Z, Qing M, et al. Graphene oxide-peptide nanocomplex as a versatile fluorescence probe of protein kinase activity based on phosphorylation protection against carboxypeptidase digestion. Analytical Chemistry. 2013;**85**:5746-5754. DOI: 10.1021/ac400336u

[6] Geißler D, Hildebrandt N. Recent developments in Förster resonance energy transfer (FRET) diagnostics using quantum dots. Analytical and Bioanalytical Chemistry. 2016;**408**:4475-4483. DOI: 10.1007/s00216-016-9434-y

[7] Ren X, Wei J, Ren J, Qiang L, Tang F, Meng X. A sensitive biosensor for the fluorescence detection of the acetylcholinesterase reaction system based on carbon dots. Colloids and Surfaces B: Biointerfaces. 2015;**125**:90-95. DOI: 10.1016/j.colsurfb.2014.11.007

[8] Lee JS, Youn YH, Kwon K, Ko NR. Recent advances in quantum dots for biomedical applications. Journal of Pharmaceutical Investigation. 2018;**48**:209-214. DOI: 10.1007/s40005-018-0387-3

[9] Lee J, Kim J, Kim S, Dal-Hee M. Biosensors based on graphene oxide and its biomedical application. Advanced Drug Delivery Reviews. 2016;**105**:275-287. DOI: 10.1016/j.addr.2016.06.001

[10] He S, Song B, Li D, Zhu C, Qi W, Wen Y, et al. A graphene nanoprobe for rapid, sensitive and multicolor fluorescent DNA analysis. Advanced Functional Materials. 2010;**20**(3);453-459. DOI: 10.1002/adfm.200901639

[11] Wimmer T, Lorenz B, Stieger K. Quantification of the vascular endothelial growth factor with a bioluminescence resonance energy transfer (BRET) based single molecule biosensor. Biosensors and Bioelectronics. 2016;**86**:609-615. DOI: 10.1016/j.bios.2016.07.058

[12] Gomes AC, Mello AL, Ribeiro MG, Garcia DG, Da Fonseca CO, Salazar MA, et al. Perillyl alcohol, a pleiotropic natural compound suitable for brain tumor therapy, targets free radicals. Archivum Immunologiae et Therapiae Experimentalis. 2017;**65**(4):285-297. DOI: 10.1007/s00005-017-0459-5

[13] Lin LJ, Grimme JM, Sun J, Lu S, Gai L, Cropek DM, et al. The antagonistic roles of PDGF and integrin avb3 in regulating ROS production at focal adhesions. Biomaterials. 2013;**34**:3807-3815. DOI: 10.1016/j.biomaterials.2013.01.092

[14] Guzy RD, Hoyos B, Robin E, Chen H, Liu L, Mansfield KD, et al. Mitochondrial complex III is required for hypoxia-induced ROS production and cellular oxygen sensing. Cell Metabolism. 2005;**1**:401-408. DOI: 10.1016/j.cmet.2005.05.001

[15] Bernardini A, Brockmeier U, Metzen E, Pfannschmidt UB, Harde E, Palmer AA, et al. Type I cell ROS kinetics under hypoxia in the intact mouse carotid body ex vivo: A FRET-based study. American Journal of Physiology: Cell Physiology. 2015;**308**:C61-C67. DOI: 10.1152/ajpcell.00370.2013

[16] Shi S, Ang EL, Zhao H. In vivo biosensors: Mechanisms, development, and applications. Journal of Industrial Microbiology & Biotechnology. 2018;**45**:491-516. DOI: 10.1007/s10295-018-2004-x

[17] Lemon CM. Optical oxygen sensing with quantum dot conjugates. Pure Applied Chemistry. 2018;**90**(9): 1359-1377. DOI: 10.1515/pac-2018-0303

[18] Lemon CM, Karnast E, Bawendi MG, Nocera DG. Two-photon oxygen sensing with quantum dot-porphyrin conjugates. Inorganic Chemistry. 2013;**52**(18):10394-10406. DOI: 10.1515/pac-2018-0303

[19] Lemon CM, Nocera DG. Comparison of self-assembled and micelle encapsulated QD chemosensor constructs for biological sensing. Faraday Discussions. 2015;**185**:249-266. DOI: 10.1039/C5FD00093A

[20] Bidmanova S, Kotlanova M, Rataj T, Damborsky J, Trtilek M, Prokop Z. Fluorescence-based biosensor for monitoring of environmental pollutants: From concept to field application. Biosensors and Bioelectronics. 2016;**84**:97-105. DOI: 10.1016/j.bios.2015.12.010

[21] Xia J, Lin M, Zuo X, Su S, Wang L, Huang W, et al. Metal ion-mediated assembly of DNA nanostructures for cascade fluorescence resonance energy transfer-based fingerprint analysis. Analytical Chemistry. 2014;**86**(14): 7084-7087. DOI: 10.1021/ac5015436

[22] Tan L, Chen Z, Zhao Y, Wei X, Li Y, Zhang C, et al. Dual channel sensor for detection and discrimination of heavy metal ions based on colorimetric and fluorescence response of the AuNPs-DNA conjugates. Biosensors and Bioelectronics. 2016;**85**:414-421. DOI: 10.1016/j.bios.2016.05.038

[23] Li H, Sun DE, Liu Y, Liu Z. An ultrasensitive homogeneous aptasensor for kanamycin based on upconversion fluorescence resonance energy transfer. Biosensors and Bioelectronics. 2014;**55**:149-156. DOI: 10.1016/j.bios.2013.11.079

[24] He X, Ma N. An overview of recent advances in quantum dots for biomedical applications. Colloids and Surfaces B: Biointerfaces. 2014;**124**(1):118-131. DOI: 10.1016/j.colsurfb.2014.06.002

[25] Luo Q, Lai J, Qiu P, Wang X. An ultrasensitive fluorescent sensor for organophosphorus pesticides detection based on RB-Ag/Au bimetallic nanoparticles. Sensors and Actuators B: Chemical. 2018;**263**:517-523. DOI: 10.1016/j.snb.2018.02.101

[26] Li X, Wu Z, Zhou X, Hu J. Colorimetric response of peptide modified gold nanoparticles: An original assay for ultrasensitive silver detection. Biosensors and Bioelectronics. 2017;**92**:496-501. DOI: 10.1016/j.bios.2016.10.075

[27] Arvand M, Mirroshandel AA. Highly-sensitive aptasensor based on fluorescence resonance energy transfer between L-cysteine capped ZnS quantum dots and graphene

oxide sheets for the determination of edifenphos fungicide. Biosensors and Bioelectronics. 2017;**96**:324-331. DOI: 10.1016/j.bios.2017.05.028

[28] Wu X, Song Y, Yan X, Zhu C, Ma Y, Du D, et al. Carbon quantum dots as fluorescence resonance energy transfer sensors for organophosphate pesticides determination. Biosensors and Bioelectronics. 2017;**94**:292-297. DOI: 10.1016/j.bios.2017.03.010

[29] Song L, Mao K, Zhou X, Hu J. A novel biosensor based on Au@Ag core-shell nanoparticles for SERS detection of arsenic(III). Talanta. 2016;**146**: 285-290. DOI: 10.1016/j.talanta. 2015.08.052

[30] Zhang Z, Zhang M, Wu X, Chang Z, Lee Y, Huy BT, et al. Upconversion fluorescence ressonance energy-transfer—A novel approach for sensitive detection of fluoroquinolones in water samples. Microchemical Journal. 2016;**124**:181-187. DOI: 10.1016/j. microc.2015.08.024

[31] Ruiz BC, Gavela AF, Herranz S, Lechuga LM. Nanophotonic label-free biosensors for environmental monitoring. Current Opinion in Biotechnology. 2017;**45**:175-183. DOI: 10.1016/j.copbio.2017.03.016

[32] Bourgeois W, Burgess JE, Stuetz RM. Online monitoring of waste-water quality: A review. Journal Chemical Technology and Biotechnology. 2001;**76**(4):337-348. DOI: 10.1002/ jctb.393

[33] Ejeian F, Etedali P, Tehrani HAM, Soozanipour A, Low ZX, Asadnia M, et al. Biosensors for waste water monitoring: A review. Biosensors and Bioelectronics. 2018;**118**:66-79. DOI: 10.1016/j.bios.2018.07.019

[34] Pasternak G, Greenman J, Ieropoulos I. Self-powered, autonomous biological oxygen demand biosensor for online water quality monitoring. Sensors and Actuators B: Chemical. 2017;**244**:815-822. DOI: 10.1016/j. snb.2017.01.019

[35] Sousa CP, Coutinho-Neto MD, Liberato MS, Kubota LT, Alves WA. Self-assembly of peptide nanostructures onto an electrode surface for nonenzymatic oxygen sensing. Journal of Physical Chemistry. 2015;**119**(2):1038-1046. DOI: 10.1021/ jp509020x

[36] Kashem MA, Suzuki M, Kimoto K, Iribe Y. An optical biochemical oxygen demand biosensor chip for enivironmental monitoring. Sensors and Actuators B: Chemical. 2015;**221**: 1594-1600. DOI: 10.1016/j.snb. 2015.07.119

[37] Zhao Q, Pan T, Xiang G, Mei Z, Jiang J, Li G, et al. Highly efficient ratiometric extracellular oxygen sensors through physical incorporation of a conjugated polymer and PtTFPP in graft copolymers. Sensors and Actuators B: Chemical. 2018;**273**:242-252. DOI: 10.1016/j.snb.2018.06.026

[38] Chu CS, Lin TH. A new portable optical sensor for dual sensing of temperature and oxygen. Sensors and Actuators B. 2014;**202**:508-515. DOI: 10.1016/j.snb.2014.05.125

[39] Mao Y, Mei Z, Wen J, Li G, Tian Y, Zhou B, et al. Honeycomb structured porous films from a platinum porphyrin-grafted poly(styrene-co-4-vinylpyridine) copolymer as an optical oxygen sensor. Sensors and Actuators B: Chemical. 2018;**257**:944-953. DOI: 10.1016/j.snb.2017.11.042

[40] Ribeiro ACC, Souza GA, Filho CPA, Martins TD. Sistema fluorescente para sensor óptico de oxigênio dissolvido, método para preparar dispositivo de sensor óptico de sistema fluorescente. Patente No. BR102015020665-A2. 2015

Chapter 7

Paper-Based Biosensors for Analysis of Water

Patrícia S. Peixoto, Ana Machado, Hélder P. Oliveira, Adriano A. Bordalo and Marcela A. Segundo

Abstract

The presence of contaminants in water generates a great concern worldwide. As contaminants, we can refer different classes of chemicals, such as pharmaceuticals, personal care products, heavy metals, and also microorganisms, such as waterborne pathogens. Some of the chemical compounds have the potential to bioaccumulate in the aquatic biota. Hence, the development of simple and portable methods for the detection of contaminants in the aquatic environment can improve their monitoring and, consequently, the study of their environmental impact. In this context, the development of paper-based analytical tools and also of biosensor devices has been exploited for quantitative and semiquantitative analysis of several contaminants in different water matrices. The association of these two analytical strategies can provide the implementation of low-cost, portable, and easily handled methods for detecting chemical and biological contaminations in water. In this chapter, we provide a review of the developed paper-based analytical biosensors, highlighting the features of the paper-based (paper substrate and fabrication procedures) and biosensor devices (transducers and biorecognition elements). Moreover, the application of the referred paper-based biosensors for the detection of different water contaminants (pathogens, pharmaceuticals, and heavy metals) in environmental and wastewater samples is discussed.

Keywords: microfluidic, paper-based devices, water analysis, water contaminants, biosensing

1. Introduction

The contamination of the different water compartments with several chemicals and by-products has become a major concern for human health and aquatic biota [1–3]. The environmental contaminants of major concern are pharmaceuticals, personal care products, pesticides and herbicides, heavy metals, and waterborne pathogens. Some of the chemical compounds are persistent to degradation and, therefore, can accumulate in aquatic organisms and sediments. Thus, the monitoring of contaminants in the aquatic environment is crucial to study their impact [4].

Currently, there is no regulation about the allowed levels of pharmaceutical compounds, including ethinylestradiol and antibiotics, in water. Concerning the maximum contaminant levels (MCL) in drinking water for the target metals presented in this work, their MCL are between 2 and 30 $\mu g\ L^{-1}$, according to the chemical species [5, 6]. Arsenic presents an MCL value of 10 $\mu g\ L^{-1}$ [5, 6], while the MCL value for mercury is 2 or 6 μg L^{-1} according to United States Environmental Protection Agency (EPA) and World Health Organization (WHO), respectively. Moreover, the MCL for lead is 15 [5] and 10 μg L^{-1} [6]. Uranium presents the higher MCL, which is 30 μg L^{-1} [5, 6]. Cadmium's maximum contaminant level corresponds to 3 and 5 μg L^{-1} according to WHO and EPA, respectively. With respect to the pathogens targeted in the paper-based biosensors, the EPA [7] recommends that *Escherichia coli* cannot exceed 126 CFU per 100 mL in fresh recreational water, while *Enterococcus* should present a maximum of 35 CFU per 100 mL in marine and freshwater.

In this context, paper-based biosensor devices combine the main features of paper substrates (cost-effectiveness, easy manipulation, and compatibility with proteins and biomolecules), with the high specificity and selectivity of the biorecognition systems of biosensors [1, 8, 9]. Furthermore, paper-based assays can be a solution in resource-limited contexts, as both sample and reagents can be introduced without any flow device, through imbibition and filtration via capillary action [10].

The first types of paper-based devices were related to semiquantitative analysis of glucose in urine and immunoassays on chromatographic paper test strips (or lateral flow) [11]. In the last decade, a new fabrication method based on wax patterning was introduced, allowing the design of well-defined channels on paper surface, which provided microfluidic features to the paper-based devices [12].

Reviews concerning the application of paper-based devices in different fields such as food, water analysis, environmental monitoring, and health diagnostics are available [8, 11, 13]. Furthermore, the application of biosensors has been extensively discussed regarding both their usefulness on assessing environmental and urban pollutants [1], and also their role as part of portable biochemical detection systems [14]. However, gathering information about the implementation of paper-based techniques coupled with biosensor devices to water analysis is still lacking. Hence, the aim of this work is to provide a description of the state of the art about the development and application of paper-based analytical biosensors to detect contaminants in water, focusing on work developed in the last 3 years.

2. Paper-based analytical devices

2.1 Substrate material

Paper is a complex material and a promising support for the development of biosensor analytical devices. Its main features, such as versatility, low-cost, and biocompatibility, generate simple and disposable bioanalytical tools using low reagent consumption (in the order of microliters) [8, 11, 12]. Paper is mainly constituted of cellulose fibers. The cellulose is a hydrophilic polymer, which makes paper substrate permeable to aqueous liquids [15].

There are different types of paper that are used for fabrication of paper-based devices (**Table 1**). Filter papers have been widely used as substrate material to paper-based devices for biosensor application [16–19]. There are a vast range of commercially available high-quality filter papers, mainly constituted of alpha cellulose, a highly stable form of cellulose. The filter papers can be classified according to different properties, such as particle retention, pore size, thickness, and flow rate.

The filter paper grade 1, considered as a medium retention and flow filter paper, has been functionalized to obtain paper-based immunosensors [16, 17]. Irvine et al. adsorbed metallothioneins in grade 1 filter paper for the detection of heavy metals [18]. In the same way, other filter paper grades have been used, such as the slow filter paper grade 42 (pore size of 2.5 μm) for the incorporation of an in vitro transcription/translation system [19].

Type of paper	Examples	Features
Cellulose filter paper	Whatman® filter paper grades 1 and 2	Hydrophilic polymer, permeable to aqueous liquids; available with different pore sizes and thickness
Cellulose chromatography paper	Whatman® chromatographic paper grades 1, and 2	Allows the concentration of nanostructures in its surface and the separation of nanoparticles in agglutination-based assays
Nitrocellulose membrane	Millipore Hi-Flow Plus HF240	Hydrophobic polymer; adequate for immobilization of complex biological structures by electrostatic interactions
Printing paper	Fabriano 5 HP paper, Boise®Aspen® 30 multiuse recycled copy paper	3D structures can be easily printed in its surface, providing microchannels or screen-printed electrodes

Table 1.
Types and features of papers used in biosensors described recently.

Cellulose chromatography paper is also an alternative as high-quality substrate in paper-based biosensors. These papers can be differentiated by their flow rate and thickness. Vijitvarasan et al. [20] developed a paper-based device taking in advantage of the separative properties of chromatographic paper in order to enhance the concentration of gold nanoparticles (AuNPs) on the surface of the paper. Thus, a lower level of reduced silver particles was detected on the surface of AuNPs. Chromatography papers have also been applied in the development of microfluidic devices based on capillary flow measurement [21, 22]. McCraken et al. [22] tested two different chromatography papers (grade 1 Chr and grade 2 Chr) to optimize the separation of immunoagglutinated particles. The chromatography paper providing the lower flow rate (grade 2, 115 mm/30 min) was selected for the immunoagglutination assay.

Derivatives of cellulose, such as nitrocellulose membranes, have been applied as substrates of paper-based devices. These membranes are naturally hydrophobic and demonstrate to be adequate for the immobilization of enzymes and proteins by electrostatic interactions [11]. For instance, Lopez-Marzo et al. [23] developed a lateral flow immunodevice with nitrocellulose membrane for the detection of Cd^{2+} in water, taking advantage of the immobilization of antibodies (2A81G5 mouse antibody and antibovine serum albumin (BSA) mouse antibody), to create zones where the probe conjugate and the positive control containing BSA would interact. A similar approach was reported for immobilization of BSA conjugate and control goat antimouse immunoglobulin for the detection of U(VI) [24].

Concerning electrochemical paper-based devices, printing paper can be applied in the development of screen-printed electrodes. Hence, carbon-based conductive ink is printed onto paper surface [25, 26]. In this context, Rengaraj et al. [27] fabricated a paper-based electrode with the high-quality printing paper using only three layers of printing.

In addition, both chromatographic paper and multiuse recycled copy paper were used for printing wax wells in paper-based devices as a confinement strategy [20], or as a low-cost alternative to microplates [28].

2.2 Fabrication procedures

There are different techniques that can be applied in order to obtain paper-based biosensor devices with variable properties, such as functionalized platforms with

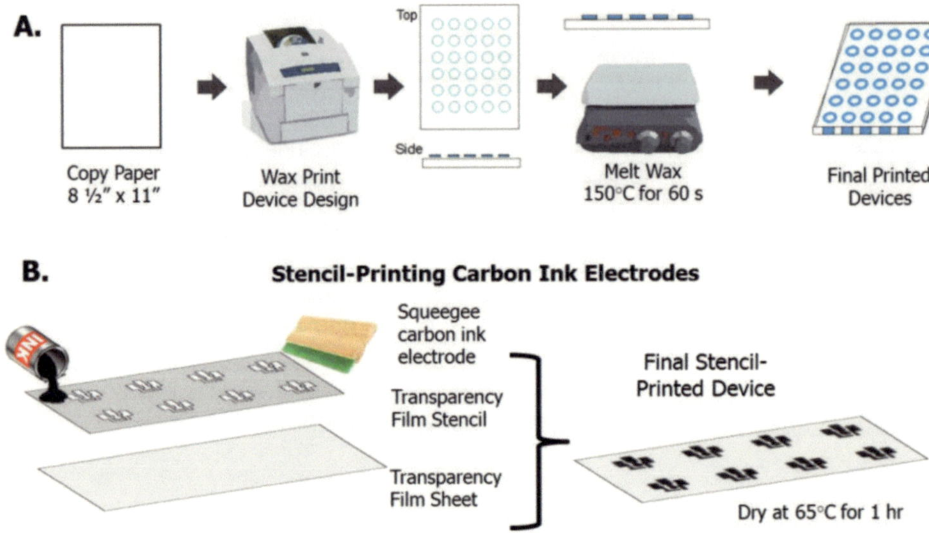

Figure 1.
Examples of fabrication schemes for production of (A) wax printed paper-based well devices and (B) stencil-printed transparency film-based carbon electrodes. Adapted and reprinted with permission from [28]. Copyright 2017 American Chemical Society.

biomolecules or cell suspensions, conductive characteristics for electrochemical analysis, and create barriers to define the reaction zones.

Wax printing is a process used to create hydrophobic barriers that define reaction microzones or fluid reservoirs [29], as exemplified in **Figure 1**. Different works [16, 17, 20, 22, 28] applied this technique using graphic design software [16, 17, 22] or stencils [28] to define the microchannel areas. The incorporation of the wax onto the microfluidic channel is performed by printing the wax onto the paper surface with subsequent heating to allow wax penetration in the paper.

Screen printing is another technique used to fabricate paper-based devices, particularly for electrochemical analysis (**Figure 1B**). For example, Rengaraj et al. [27] fabricated a paper-based electrode by printing three layers of a carbon-based conductive ink onto hydrophobic printer paper. Other fabrication techniques include a simple procedure of cutting by punching [18, 19], obtaining discs with millimetric dimensions that can be functionalized and/or introduced into devices, such as commercial screen-printed electrodes [18].

Furthermore, lateral flow immune-based devices can be fabricated by assembling different layers, which include the conjugation of pad strip (signal producer), the nitrocellulose membrane, as well as the sample and absorption pad [23, 24]. Stocker et al. [30] applied a simple technique based on premarking the spots with a pencil, with subsequent physical deposition of a cell suspension and drying of the paper strips.

3. Integrated biosensors methods

3.1 Transducers

Biosensors can be defined as analytical devices, which integrate or associate a biorecognition element and a transducer. The bioelement recognizes the target analyte and the transducer converts the biochemical interaction to a measurable signal [1]. The most frequently applied transducers are based on optical, electrochemical, thermal, and piezoelectric properties. This work focused on the transducer types most used in the paper-based biosensors for analysis of water (electrochemical, optical, and piezoelectric).

The electrochemical paper-based biosensors are based on the modification of paper-based platforms placed in commercial screen-printed electrodes [17, 18], or rely on the fabrication of lab-made functionalized paper-based screen-printed electrodes [27].

Concerning the optical approaches used in the paper-based devices, the most applied strategies were based on colorimetric measurement [19, 20, 28] using image processing algorithms as an analytical system. In this context, a colorimetric based approach for the detection of Pb^{2+} and for U(VI) resorted to the acquisition of images of the paper spots with a digital camera, and images processed using the ImageJ software [20, 24]. On the other hand, Adkins et al. [28] developed a paper-based colorimetric method for the detection of *Escherichia coli*, which was based on a smartphone for image acquisition and ImageJ software for image processing.

Other example of mobile-based strategy involves the quantification of different pathogens (*E. coli* and Zika virus) as a function of capillary flow rate using a smartphone as a photometric detector [21]. This photometric approach was based on video recording of an immunoassay followed by comparison of the capillary flows between different analyte concentrations. Other colorimetric method was based on the detection of Cd^{2+} [23] in drinking water with a lateral flow immunosensor device. The measurement of color intensity was performed with COZART™ RapidScan color intensity portable reader. Colorimetry in paper-based biosensor devices was developed as a semiquantitative approach for the detection of arsenite [30]. This method was based on a bacterial biosensor deposited onto a paper strip. The developed color measurement was performed by comparison with spots containing known arsenite concentrations. Furthermore, a method based on fluorescence was applied as transducer for the detection of ethinylestradiol in a paper-based immunoassay [16]. For this, a LED-based system was constructed and used as excitation source. The fluorescence emission was measured with a scientific-grade spectrometer.

A piezoelectric strategy was also implemented for measurement of immunoagglutinated samples for the detection of *E. coli* and Zika virus [22]. This approach was based on particle rheology of the immunoagglutinated samples. In order to monitor the movement of the suspension of particles in a microfluidic paper-based platform, videos were taken with a smartphone and flow distance was measured every five frames.

3.2 Biorecognition elements

The biorecognition element has a strong and selective affinity to the target. There are several types of biorecognition elements, such as natural biomolecules (nucleic acids, antibodies, enzymes, and other proteins), synthetic bioelements (molecularly imprinted polymers, aptamers), or whole cells (**Figure 2**) [1].

Different types of biorecognition elements have been applied for the development of paper-based biosensors for the detection of target analytes in water. An in vitro transcription/translation system reconstituted from purified recombinant components necessary for *E. coli* translation of β-galactosidase enzyme was immobilized on paper as a turn on/turn off switcher for the presence of antibiotics inhibiting bacterial protein synthesis [19].

Concerning the application of antibodies as biorecognition elements, specific antibodies (polyclonal rabbit anti-EE2) have been applied for the detection of the estrogen ethinylestradiol in river water samples [16, 17]. In addition, suspensions of antibody-conjugated particles were used for the detection of two target pathogens (*E. coli* K12 and Zika virus) [21]. For the detection of U(VI), immobilized U(VI)-2,9-dicarboxyl-1,10-phenanthroline-BSA conjugate worked as a competitive probe

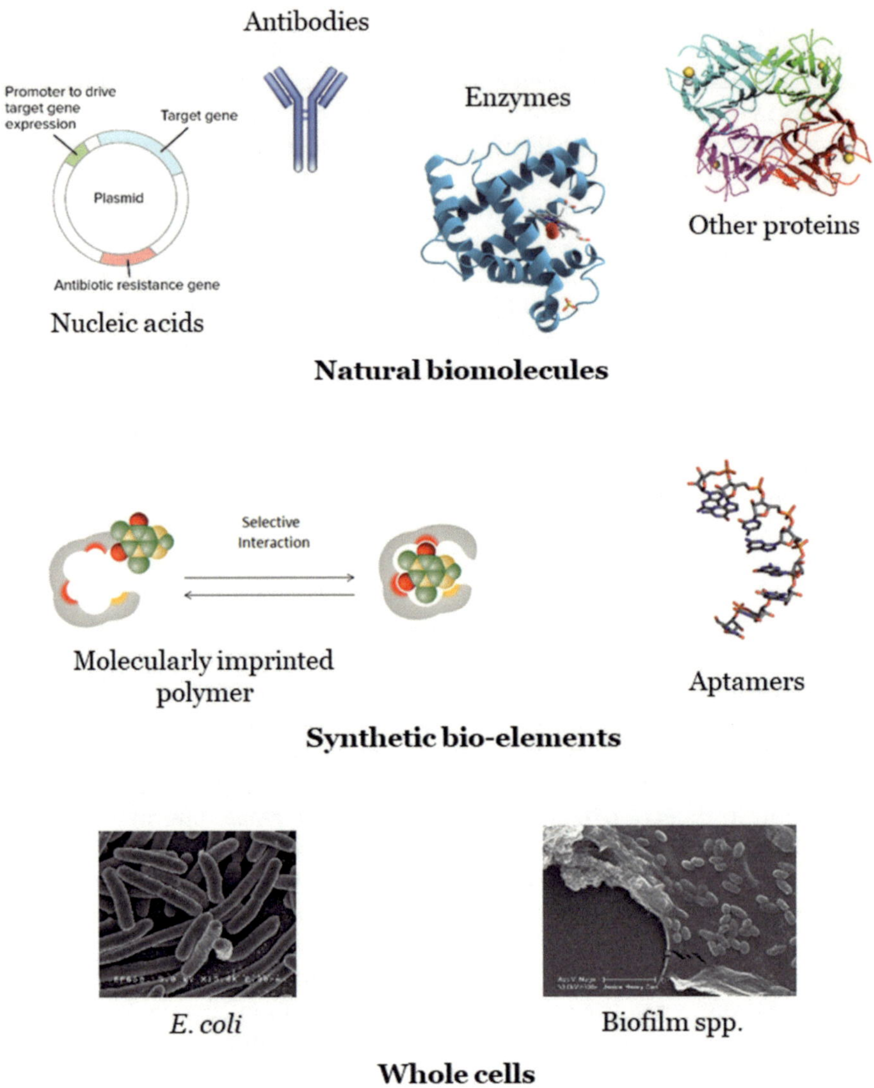

Figure 2.
Examples of biorecognition elements found in paper-based biosensors.

for the antibody 12F6-AuNP conjugate, as the antibody 12F6 has an increased affinity to U(VI)-2,9-dicarboxyl-1,10-phenanthroline complex [24].

Biomolecules aiming the detection of toxic metals can also be referred. A complex comprising magnetic beads, gold nanoparticles (AuNPs), and the functional nucleotide GR5-DNAzyme was applied as a biorecognition element of lead ion [20]. In another work, the recombinant human metallothionein 1a, a metal-binding protein [18], was used for the recognition of As^{3+} and Hg^{2+} in water. The tetrameric protein lectin concanavalin A (obtained from *Canavalia ensiformis*), selective to carbohydrates on bacterial cells, was selected as a biorecognition element of bacterial cultures from sewage sludge [27].

Finally, whole-cell living bacterial biosensors for arsenite detection were based on genetically engineered *E. coli*, where the *ars* operon (set of structural and regulatory genes whose expression is controlled through arsenite binding) was modified with a sequence for expression of β-galactosidase as a reporter protein in the presence of the target analyte [30]. Whole cells (biofilm formed from anaerobic sludge) were also employed in the biosensor proposed by Chouler et al. [25] for the assessment of toxic compounds in water in a microbial fuel cell device. The detection principle was based on the conversion of the chemical energy contained in organic matter into electricity via the metabolic processes of microorganisms. Hence, a microbial biofilm is placed on the anode surface, where the electroactive bacteria mediate the transference of electrons to the electrode upon their metabolic activity. Any factor disrupting this (water pollution for instance) will disrupt this signal. A similar approach was proposed by Xu et al. [26] using a wastewater bacteria consortium.

4. Applications

In this section, the application of paper-based biosensors for the detection of different types of target analytes in water samples is discussed. In **Table 2**, the main features of the target analyte, the sample type, the paper substrate, and the fabrication method of the paper-based device, the method of detection, and the biorecognition element are summarized.

Pharmaceuticals are among the targets. Indeed, they are considered emerging environmental contaminants, as they can be harmful to human health and to aquatic life. In this context, the synthetic hormone—ethinylestradiol, one of the main compounds of oral contraceptives, is considered an emerging pollutant due to its potential high estrogenic effect on the biota. Scala-Benuzzi et al. developed two different methods for the detection of ethinylestradiol in river water samples using an antiethinylestradiol specific antibody [16, 17]. In both approaches, the water samples were filtered, and pH was adjusted to 7.0 with phosphate buffer before the analysis. In one work, a fluorescent paper-based biosensor was implemented [16]. This methodology presented a limit of detection (LOD) of 0.05 ng L^{-1}, which is mainly related to the high sensitivity of fluorescence methods. In another work, ethinylestradiol was detected in river water with a paper-based immunosensor based on electrochemical analysis [17], which also reached a low LOD value of 0.1 ng L^{-1}, suitable for environmental analysis.

Antibiotics are another group of pollutants of great concern due to the global threat of antimicrobial resistance and the excessive, and sometimes abusive, use of these compounds. A colorimetric biosensor for screening of several antibiotics (paromomycin, tetracycline, chloramphenicol, and erythromycin) inhibiting bacterial protein synthesis was applied for the detection of antibiotics in surface water [19]. The method was based on the ability of these antimicrobials to inhibit β-galactosidase synthesis. When a water sample without the target antibiotics was placed in the paper-based device, the enzyme β-galactosidase was synthetized and its activity induced a color change on the paper disc surface. However, when antibiotics were present, the inhibition of β-galactosidase synthesis prevented the change of color. Despite the limit of detection was on the microgram per milliliter level (0.5, 2.1, 0.8, and 6.1 μg mL^{-1} for paromomycin, tetracycline, chloramphenicol, and erythromycin, respectively), this biosensor can be applied as a simple and portable screening methodology.

Heavy metals are naturally present in the environment. However, these elements can be toxic to human and aquatic organisms even at low concentrations. Moreover, their presence can be increased by industrial and agriculture activities. Vijitvarasan et al. implemented a paper-based scanometric biosensor for the detection of lead in water [20]. The biosensor was applied to river water samples. These samples were filtered, diluted 10 times with 10 mM tris-acetate buffer and spiked with different Pb^{2+} concentrations before analysis. An LOD value of 0.9 nM was determined. Furthermore, a method using a sensitive gold nanoparticle-based lateral flow immunodevice [23] was applied for the quantification of cadmium. Drinking water

Target	Sample type	Paper substrate; fabrication method	Detection method	Biorecognition element	LOD	Reference
Pharmaceuticals						
Ethinylestradiol (EE2)	River water	Filter paper grade 1; wax printing	F	Antibody (anti-EE2)	0.05 ng L^{-1}	[16]
EE2	River water	Filter paper grade 1; wax printing	E	Antibody (anti-EE2)	0.1 ng L^{-1}	[17]
Antibiotics inhibiting protein synthesis[a]	Surface water	Filter paper discs, 1442-055; cutting by punching	C	Enzyme (β-galactosidase)	0.5–6.1 μg L^{-1}	[19]
Metals						
Arsenic, mercury	Ultrapure water	Filter paper grade 1; cutting by punching	E	Recombinant human metallothionein 1a	13 ppb (As^{3+}); 45 ppb (Hg^{2+})	[18]
Lead ion	River water; synthetic urine	Chromatography paper; wax printing	C	GR5-DNAzyme	0.3 nM	[20]
Uranium (VI)	River water	Cellulose and nitrocellulose membranes; assembling of different layers in a plastic backing card	C	12F6 antibody against U(VI)-chelator complex	36 nM	[20, 24]
Arsenite	Ground-water	0.5 × 4 cm paper; cutting	C	*E. coli*	8 μg L^{-1}	[30]
Cd^{2+}	Drinking water, tap water	Hi-Flow Plus nitrocellulose membrane; assembling of different layers in a plastic backing card	C	Cd–EDTA–BSA–AuNP	0.1 ppb	[23]
Pathogens						
E. coli and Zika virus	Deionized water; serum simulant	Paper Chr. grade 2; wax printing	P	Antibody-conjugated particles	1 log CFU mL^{-1} [b]; 20 pg mL^{-1} [c]	[21]
E. coli K12 and Zika virus	Deionized water; simulated serum	Paper Chr. grade 2; wax printing	R	Antibody (goat polyclonal)	2 log CFU mL^{-1} [b]; 0.531[d]	[22]
Bacterial cultures from sewage sludge	Synthetic wastewater	Cotton-based paper; screen printing	E	Lectin concanavalin A	1.9×10^{3} CFU mL^{-1}	[27]

Target	Sample type	Paper substrate; fabrication method	Detection method	Biorecognition element	LOD	Reference
E. coli, *Enterococcus* spp.	Lagoon water, alfalfa sprout	Multiuse recycled copy paper; wax printing	C	Substrates ONP and PNP	81 μM (ONP), 119 μM (PNP)	[28]
Others						
Water toxicity	Artificial wastewater	Cotton-based paper; screen printing	E	Biofilm formed from anaerobic sludge	0.1[e]	[25]
Shock pollution	Wastewater	Filter paper; ink coating	E	Bacteria consortium	0.022[f]	[26]

F, fluorescence; E, electrochemical; C, colorimetric; P, photometric; R, rheology-based measurement. PNP: p-nitrophenol; ONP: o-nitrophenol.

[a]Paromomycin, tetracycline, chloramphenicol, and erythromycin.

[b]LOD value for E. coli.

[c]LOD value for Zika virus.

[d]LOD value for Zika virus expressed as transcription copies mL^{-1}.

[e]Formaldehyde concentration monitored (% v/v).
[f]Power output slope for Cr (VI).

Table 2.
Paper-based biosensors for analysis of pharmaceuticals, heavy metals, and waterborne pathogens in water.

samples were spiked with 10 and 100 ppb of Cd^{2+}, and other 11 metals commonly found in such type of water, containing also EDTA and ovoalbumin (masking agent). An LOD of 0.1 ppb was achieved.

A semiquantitative approach based on a paper-based bacterial biosensor was applied [30] for the detection of arsenite in groundwater samples. It was observed that arsenite produced a visible blue color from substrate of β-galactosidase (reporter protein) at arsenite concentration above 8 $\mu g\ L^{-1}$. Finally, a paper-based lateral flow device was developed for uranium (VI) determination with an LOD (36 nM) below the action level established by the World Health Organization (126 nM) using an immunological competitive approach [24]. These sensors are a suitable tool for field analysis, in opposition to conventional time-consuming and expensive techniques performed under lab environment.

Sensors based on microbial metabolism were developed for application in wastewaters. Pollution peaks, meaning the abrupt change in concentration of organic and metal pollutant in wastewater treatment plants, can compromise the biological treatment phases by killing or inhibiting microorganisms present in sludge. Hence, untargeted sensors were developed using either biofilms [25] or bacteria consortium [26] to report spiking pollution in wastewater influents.

Waterborne pathogens are a major public health as they can lead to several diseases such as cholera, typhoid fever, and dysentery. Hence, accessible, cheap, and disposable analytical tools for monitoring the presence of these pathogens are mandatory, especially in areas with low resources. In this context, an electrochemical paper-based biosensor [28] was developed for the detection of *E. coli* in water, as an indicator of fecal contamination and an indirect indicator of the presumptive presence of other gastrointestinal bacteria. Different *E. coli* strains (both pathogenic and nonpathogenic) were detected in uninoculated and inoculated lagoon water. The method was able to detect as low as 10 CFU mL^{-1} of pathogenic and nonpathogenic *E. coli*.

5. Conclusions

The paper-based biosensors developed for quantification of the synthetic hormone EE2 presented higher sensitivity when compared to the more complex and

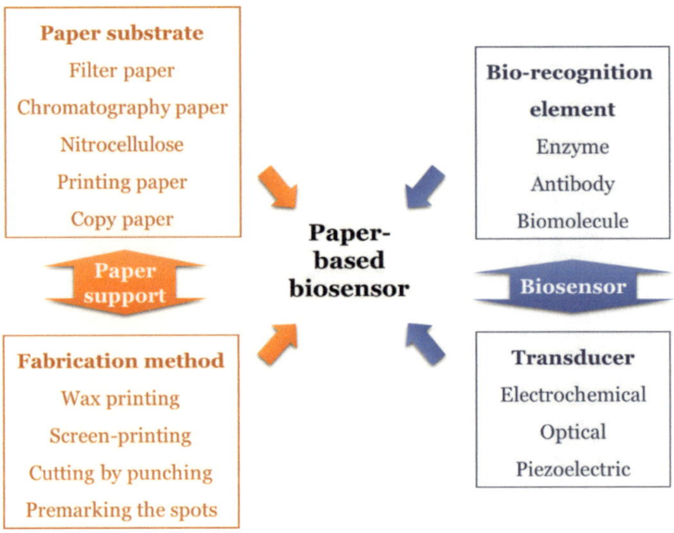

Figure 3.
Schematic representation of the elements that compose paper-based biosensors.

expensive LC-MS/MS methods [31, 32]. With respect to the developed biosensors for quantification of different metals (uranium [24], arsenite [30], and cadmium [23]), the LOD was lower than their respective MCLs, thus complying with regulatory requirements. Furthermore, *E. coli* was detected [21, 22] at concentration level similar or lower than the maximum CFU/mL allowed in fresh recreational water. Hence, the use of paper-based platforms in biosensors has allowed the development of simple, specific, sensitive, and portable devices for the detection of several types of target analytes in water, with possible features summarized in **Figure 3**. Most of the reported methods were applied to surface water and drinking water samples only, which are samples containing a reduced amount of organic matter when compared to wastewater. Hence, efforts to develop sensors that can deal with more complex matrices should be pursued, encompassing strategies that accommodate sample pretreatment.

The most frequently used transducers comprised electrochemical and optical methods, with analytical strategies based on colorimetric reactions, associated with image processing analysis. Besides the recent advances in the development of paper-based analytical tools and biosensor devices, their association to the analysis of contaminants in water is still an open research field with a high potential for the implementation of new portable and low-cost analytical methods for *in situ* analysis.

Acknowledgements

This work received financial support from the European Union (FEDER funds through COMPETE POCI-01-0145-FEDER-031756) and National Funds (FCT/MEC, Fundação para a Ciência e Tecnologia and Ministério da Educação e Ciência) through project PTDC/CTAAMB/31756/2017.

Conflict of interest

We declare that there is no conflict of interest.

Author details

Patrícia S. Peixoto[1], Ana Machado[2,3], Hélder P. Oliveira[4], Adriano A. Bordalo[2,3] and Marcela A. Segundo[1]*

1 LAQV, REQUIMTE, Department of Chemical Sciences, Faculty of Pharmacy, University of Porto, Portugal

2 ICBAS, Institute of Biomedical Sciences Abel Salazar, University of Porto, Portugal

3 CIIMAR/CIMAR, Interdisciplinary Centre of Marine and Environmental Research, University of Porto, Portugal

4 INESC TEC, Faculty of Sciences, University of Porto, Portugal

*Address all correspondence to: msegundo@ff.up.pt

References

[1] Hashemi Goradel N, Mirzaei H, Sahebkar A, Poursadeghiyan M, Masoudifar A, Malekshahi ZV, et al. Biosensors for the detection of environmental and urban pollutions. Journal of Cellular Biochemistry. 2018;**119**(1):207-212. DOI: 10.1002/jcb.26030

[2] Ebele AJ, Abou-Elwafa Abdallah M, Harrad S. Pharmaceuticals and personal care products (PPCPs) in the freshwater aquatic environment. Emerging Contaminants. 2017;**3**(1):1-16. DOI: 10.1016/j.emcon.2016.12.004

[3] Peixoto PS, Toth IV, Segundo MA, Lima JLFC. Fluoroquinolones and sulfonamides: Features of their determination in water. A review. International Journal of Environmental Analytical Chemistry. 2016;**96**(2):185-202. DOI: 10.1080/03067319.2015.1128539

[4] Zulkifli SN, Rahim HA, Lau WJ. Detection of contaminants in water supply: A review on state-of-the-art monitoring technologies and their applications. Sensors and Actuators B: Chemical. 2018;**255**:2657-2689. DOI: 10.1016/j.snb.2017.078

[5] EPA. Edition of the Drinking Water Standards and Health Advisories. EPA 822-F-18-001. Washington, DC: U.S. EPA Office of Water; 2018

[6] World Health Organization. Guidelines for Drinking-Water Quality: Fourth Edition Incorporating First Addendum. 4th ed + 1st add. World Health Organization; 2017

[7] EPA. Recreational Water Quality Criteria. U.S. EPA Office of Water 820-F-12-058; 2012

[8] Busa LSA, Mohammadi S, Maeki M, Ishida A, Tani H, Tokeshi M. Advances in microfluidic paper-based analytical devices for food and water analysis. Micromachines. 2016;7(5):-21. DOI: 10.3390/mi7050086

[9] Ejeian F, Etedali P, Mansouri-Tehrani HA, Soozanipour A, Low ZX, Asadnia M, et al. Biosensors for wastewater monitoring: A review. Biosensors & Bioelectronics. 2018;**118**:66-79. DOI: 10.1016/j.bios.2018.07.019

[10] Aydindogan E, Celik EG, Timur S. Paper-based analytical methods for smartphone sensing with functional nanoparticles: Bridges from smart surfaces to Global Health. Analytical Chemistry. 2018;**90**:12325-12333. DOI: 10.1021/acs.analchem.8b03120

[11] Liana DD, Raguse B, Gooding JJ, Chow E. Recent advances in paper-based sensors. Sensors. 2012;**12**(9):11505-11526. DOI: 10.3390/s120911505

[12] Martinez AW, Phillips ST, Butte MJ, Whitesides GM. Patterned paper as a platform for inexpensive, low-volume, portable bioassays. Angewandte Chemie International Edition. 2007;**46**(8): 1318-1320. DOI: 10.1002/anie.200603817

[13] Almeida MIGS, Jayawardane BM, Kolev SD, McKelvie ID. Developments of microfluidic paper-based analytical devices (muPADs) for water analysis: A review. Talanta. 2018;**177**:176-190. DOI: 10.1016/j.talanta.2017.08.072

[14] Zhang D, Liu Q. Biosensors and bioelectronics on smartphone for portable biochemical detection. Biosensors & Bioelectronics. 2016;**75**:273-284. DOI: 10.1016/j.bios.2015.08.037

[15] Li X, Tian JF, Shen W. Progress in patterned paper sizing for fabrication of paper-based microfluidic sensors. Cellulose. 2010;**17**(3):649-659. DOI: 10.1007/s10570-010-9401-2

[16] Scala-Benuzzi ML, Takara EA, Alderete M, Soler-Illia G, Schneider RJ, Raba J, et al. Ethinylestradiol quantification in drinking water sources using a fluorescent paper based immunosensor. Microchemical Journal. 2018;**141**:287-293. DOI: 10.1016/j.microc.2018.05.038

[17] Scala-Benuzzi ML, Raba J, Soler-Illia G, Schneider RJ, Messina GA. Novel electrochemical paper-based immunocapture assay for the quantitative determination of ethinylestradiol in water samples. Analytical Chemistry. 2018;**90**(6):4104-4111. DOI: 10.1021/acs.analchem.8b00028

[18] Irvine GW, Tan SN, Stillman MJ. A simple metallothionein-based biosensor for enhanced detection of arsenic and mercury. Biosensors-Basel. 2017;**7**(1):11. DOI: 10.3390/bios7010014

[19] Duyen TTM, Matsuura H, Ujiie K, Muraoka M, Harada K, Hirata K. Paper-based colorimetric biosensor for antibiotics inhibiting bacterial protein synthesis. Journal of Bioscience and Bioengineering. 2017;**123**(1):96-100. DOI: 10.1016/j.jbiosc.2016.07.015

[20] Vijitvarasan P, Oaew S, Surareungchai W. Paper-based scanometric assay for lead ion detection using DNAzyme. Analytica Chimica Acta. 2015;**896**:152-159. DOI: 10.1016/j.aca.2015.09.011

[21] Klug KE, Reynolds KA, Yoon JY. A capillary flow dynamics-based sensing modality for direct environmental pathogen monitoring. Chemistry: A European Journal. 2018;**24**(23):6025-6029. DOI: 10.1002/chem.201800085

[22] McCracken KE, Tat T, Paz V, Reynolds KA, Yoon JY. Immunoagglutinated particle rheology sensing on a microfluidic paper-based analytical device for pathogen detection. In: ASABE Annual International Meeting Spokane. United States: American Society of Agricultural and Biological Engineers; 2017

[23] Lopez-Marzo AM, Pons J, Blake DA, Merkoci A. High sensitive gold-nanoparticle based lateral flow immunodevice for Cd^{2+} detection in drinking waters. Biosensors & Bioelectronics. 2013;**47**:190-198. DOI: 10.1016/j.bios.2013.02.031

[24] Quesada-Gonzalez D, Jairo GA, Blake RC, Blake DA, Merkoci A. Uranium (VI) detection in groundwater using a gold nanoparticle/paper-based lateral flow device. Scientific Reports. 2018;**8**. DOI: 10.1038/s41598-018-34610-5

[25] Chouler J, Cruz-Izquierdo A, Rengaraj S, Scott JL, Di Lorenzo M. A screen-printed paper microbial fuel cell biosensor for detection of toxic compounds in water. Biosensors & Bioelectronics. 2018;**102**:49-56. DOI: 10.1016/j.bios.2017.11.018

[26] Xu ZH, Liu YC, Williams I, Li Y, Qian FY, Zhang H, et al. Disposable self-support paper-based multi-anode microbial fuel cell (PMMFC) integrated with power management system (PMS) as the real time "shock" biosensor for wastewater. Biosensors & Bioelectronics. 2016;**85**:232-239. DOI: 10.1016/j.bios.2016.05.018

[27] Rengaraj S, Cruz-Izquierdo A, Scott JL, Di Lorenzo M. Impedimetric paper-based biosensor for the detection of bacterial contamination in water. Sensors and Actuators B: Chemical. 2018;**265**: 50-58. DOI: 10.1016/j.snb.2018.03.020

[28] Adkins JA, Boehle K, Friend C, Chamberlain B, Bisha B, Henry CS. Colorimetric and electrochemical bacteria detection using printed paper- and transparency-based analytic devices. Analytical Chemistry.

2017;**89**(6):3613-3621. DOI: 10.1021/acs.analchem.6b05009

[29] Carrilho E, Martinez AW, Whitesides GM. Understanding wax printing: A simple micropatterning process for paper-based microfluidics. Analytical Chemistry. 2009;**81**(16): 7091-7095. DOI: 10.1021/ac901071p

[30] Stocker J, Balluch D, Gsell M, Harms H, Feliciano J, Daunert S, et al. Development of a set of simple bacterial biosensors for quantitative and rapid measurements of arsenite and arsenate in potable water. Environmental Science & Technology. 2003;**37**(20):4743-4750. DOI: 10.1021/es034258b

[31] Viglino L, Aboulfadl K, Prevost M, Sauve S. Analysis of natural and synthetic estrogenic endocrine disruptors in environmental waters using online preconcentration coupled with LC-APPI-MS/MS. Talanta. 2008;**76**(5):1088-1096. DOI: 10.1016/j.talanta.2008.05.008

[32] Avar P, Zrinyi Z, Maasz G, Takatsy A, Lovas S, G-Toth L, et al. Beta-estradiol and ethinyl-estradiol contamination in the rivers of the Carpathian Basin. Environmental Science and Pollution Research. 2016;**23**(12):11630-11638. DOI: 10.1007/s11356-016-6276-2

Chapter 8

The Use of Biosensors for Biomonitoring Environmental Metal Pollution

Efraín Tovar-Sánchez, Ramón Suarez-Rodríguez, Augusto Ramírez-Trujillo, Leticia Valencia-Cuevas, Isela Hernández-Plata and Patricia Mussali-Galante

Abstract

The use of biosensors for biomonitoring environmental health has gained much attention in the last decades. The environment is continuously loaded with xenobiotics released by anthropogenic activities that pollute ecosystems, putting their integrity at risk. Therefore, there is an urgent need to study the negative effects of xenobiotics, specifically chemical agents. Biosensors or organisms that integrate exposure to pollutants in their environment and which respond in some measurable and predictable way are useful tools to study the extent of chemical pollution and its consequences across levels of biological organization. Among chemical pollutants, heavy metals are among the most toxic elements to nearly all living organisms. Wildlife is chronically exposed to complex metal mixtures in which effects on ecosystem health are difficult to assess. Therefore, different organisms may serve as biosensors to estimate detrimental effects of metal pollution. In this chapter, we will analyze bacteria, small mammals, some plant species, and lichens as biosensors for environmental metal pollution. Also, we will assess the importance of using different biomarkers on biosensors.

Keywords: heavy metals, environmental health, bacteria, small mammals, plants, lichens

1. Introduction

Human activity generates increasing amounts of new compounds that are released into the environment without prior knowledge of their potential toxicity or impact in living organisms. Heavy metals (HM) and metalloid such as arsenic (As), cadmium (Cd), mercury (Hg), lead (Pb), and aluminum (Al) are major environmental pollutants, particularly in industrial areas. Heavy metals are generated as a result of anthropogenic activities such as metal-working industries, cement factories, mining industry, smelting plants, refineries, and traffic and heating systems [1]. HM and their ions are ubiquitous and by definition are metals having atomic weights between 63.5 and 200.6 g mol^{-1} and specific gravity greater than 5 g cm^{-3} [2]. Living organisms require small doses of some essential HM, including cobalt (Co), copper (Cu), iron (Fe), manganese (Mn), molybdenum (Mo), vanadium (V), strontium (Sr), and zinc (Zn). However, in the case of essential metals and very toxic metals, excessive levels and, respectively, even small doses influence both the ecosystem and human health [3]. Nonessential HM which affect the surface water systems are Cd, chromium (Cr), Hg, Pb, As, and antimony (Sb).

Some metals have been classified as toxic, persistent, and accumulative elements. According to the Agency for Toxic Substances and Disease Registry [4], among the 10 most hazardous substances to human health, four are toxic metals: Pb, Hg, As, and Cd. Interestingly, Pb and Hg rank first among the most harmful metals to plants, followed by Cu, Cd/As, Co/Ni, and Mn [5]. Pb and Hg have been reported as mutagenic agents in plants [6]. Due to its chemical similarity to phosphorous, arsenic may interfere with several physiological and biochemical processes [6]. Cd does not appear to have any physiological function, with the exception of the marine diatom *Thalassiosira weissflogii* that possesses a carbonic anhydrase with Cd as its metal center [7]. Cu, Ni, Co, Zn, and Mn are all plant micronutrients. They participate in prosthetic groups and as cofactors of many proteins and are therefore essential for growth and development and, however, at high concentrations cause oxidative stress [8]. Al is another toxic element with significant implications for agriculture, because 30% of the world's land areas consist of acid soils [9].

Exposure to toxic metals can result in inhibition of seed germination, photosynthesis, and plant growth and consequently causes yield losses. These symptoms are normally related with overproduction or reactive oxygen species (ROS), changes in the permeability and structure of cell membranes, imbalance of mineral nutrients, incorporation of the metal into S-containing molecules, and cell cycle disruption [10]. Also, environmental metal exposure can affect all levels of biological organization. HM bioaccumulation in plants, lichens, small mammals, and bacteria might respond to this chemical stress on the molecular, cellular, or morphological scale and, at population, community and even ecosystem levels [11]. These different types of responses to toxic stress induced by HM (biomarkers) offer a powerful tool for documenting the extent of exposure and the effects of environmental metal contamination [12], revealing the potential use of plant, lichens, small mammals, and bacterial species as biosensors.

2. The use of plants as biosensors of heavy metal pollution

The use of plants as biosensors has a long history. For decades, they were used as a part of ecological risk assessment of agricultural and industrial chemicals, solid wastes, food additives, and chemically and radioactively polluted soil and water. However, the use of plants as environmental biomonitors has some drawbacks. Despite being higher eukaryotes, plants have completely different mechanisms of uptake, distribution, storage, compartmentalization, and metabolism of various pollutants [13].

2.1 Plants as biosensors

Vegetal species have the ability to absorb metals, particularly those essential for their development and growth through their root system from the soil, water, and overground vegetative organs from the atmosphere [14]. Also, these chemicals may be transported, transformed, stored, and accumulated in different cells and plant tissues. Genotoxicity or DNA damage is an early effect biomarker at the molecular level that has been used in several plant species exposed to HM. Metal binding to the cell nucleus causes damage including DNA base modifications, inter- and intramolecular cross-linkage of DNA and proteins, DNA strand breaks, rearrangements, and de-purination [15]. The use of DNA strand breaks as biomarker of genotoxicity is common in plants, for example, *Bacopa monnieri* (Plantaginaceae) exposed to Cd [16]; *Nicotiana tabacum* and *Solanum tuberosum* (Asteraceae) exposed to Cd, Cu, Pb, and Zn [17]; *Pisum sativum* (Fabaceae) exposed to Cr (VI) [18]; *Lycopersicum esculentum* (Solanaceae) and *Cucumis sativus* (Cucurbitaceae) exposed to Cu [19]; and *Prosopis laevigata* (Fabaceae [20]), *Vachellia farnesiana* (Fabaceae [21]), *Pithecellobium dulce* [22], and *Wigandia urens* exposed to Pb, Cu, and Zn [23]. These alterations at the DNA level can trigger changes at the biochemical level that lead to diverse effects at the cellular, physiological, or morphological level, as early effects of exposure to HM in plants [18, 24].

Among the responses at the cellular level that occur in plants as a result of exposure to HM, we can mention oxidative damage, the production of chelating agents, and alterations in cell division [25]. In the case of oxidative damage, reactive oxygen species (ROS) production indirectly influences the production of antioxidant enzymes [26]. For example, the hydrogen peroxide (H_2O_2) acts as a signaling molecule in response to HM and other stresses [27]. Under a pollution scenario by HM, H_2O_2 levels increase in response to Cu and Cd treatment as it has been reported in *Arabidopsis thaliana* [28], in Hg exposure in tomato (*Lycopersicon esculentum* [29]), and in response to Mn toxicity in barley (*Hordeum vulgare* [29]). As a consequence of oxidative stress, the plants experience cellular damage and accumulate metal ions that disturb cellular ionic homeostasis [30]. To minimize these detrimental effects, plants have evolved detoxification mechanisms based on chelation and subcellular compartmentalization. Chelation of HM is a detoxification strategy and the best characterized classes of HM chelator in plants are phytochelatins (PCs) and metallothioneins (MTs) [25, 30, 31].

PCs are capable of chelating HM, thereby reducing the concentration of cytotoxic free metal ions [32]. In particular, the synthesis of this type of chelant proteins is quickly active with the presence of Cd, Cu, Zn, Ag, Au, Hg, and Pb [33]. For example, Cd is not essential for plant growth, but it is readily taken up by many plant species. Higher plants react to excess Cd by stimulating sulfate absorption [34] and production of PCs involved in Cd chelation and transport into vacuoles [35, 36]. Increased availability of Cd for root uptake may cause considerable alterations in mineral nutrition [37, 38], lipid biosynthesis [39], photosynthetic rate [40], and nitrogen metabolism [41] in plants. Consequently, this led to a severe growth inhibition [42, 43] and finally death [44]. PC induction has been reported in copper-tolerant plants of *Mimulus guttatus* (Phrymaceae [45]), *Brassica juncea* (Brassicaceae) following the intracellular accumulation of Cd [46], *Rauwolfia serpentina* (Apocynaceae), *Arabidopsis thaliana* (Brassicaceae), and *Silene vulgaris* (Caryophyllaceae) exposed to As [47] and *Lotus japonicus* (Fabaceae) exposed to Cd [48].

Metallothioneins (MTs) are proteins that play a key role in the binding and transport of various metals [49]. The structure of these highly conserved proteins is linked to their role in the homeostasis of essential metals such as Zn and Cu and detoxification of toxic elements such as Cd and Hg [50]. In wheat (*Triticum*) and in rice (*Oryza sativa*), MTs are induced by metal ions, such as Cu and Cd; in *A. thaliana* MT gene expression is activated in response to Cu and Cd [51]. MTs bind metal ions in *Cicer arietinum* (Fabaceae), *Quercus suber* (Fagaceae), and *Triticum aestivum* (Poaceae) exposed to Zn and Cd [52, 53].

Also, HM mixtures affect cell division [54]. In general, Pb, Cd, Fe, and Zn reduce the synthesis of the cell wall components, causing damage to the Golgi apparatus and other cell organelles. The inhibition of mitosis is also limited by links between HM and cell wall pectin, which becomes more rigid and limits both the expansion and size of the intracellular space [55]. For example, Lerda found that Pb reduces the frequency of mitotic cells and increases the frequency of aberrant cells in onion (*Allium cepa*) [56]. Also, in corn plants (*Zea mays*), Eun and colleagues found that HM intervene in cellular division affecting the microtubules, which produces weakness in the cellular structure and the formation of binucleate cells in metaphase [57]. The aforementioned effects have been related mainly with the Pb and their synergies with other metals such as Al and Cu.

Likewise, HM interfere with ionic homeostasis and enzyme activity, resulting in physiological alterations which involve single organs (such as nutrient uptake by the roots) followed by more general processes such as germination, growth, photosynthesis, plant water balance, primary metabolism, and reproduction [30, 58]. Indeed, visible symptoms of heavy metal toxicity include chlorosis, leaf rolling and necrosis, senescence, wilting and stunted growth, low biomass production, limited numbers of seeds, and eventually death [58]. For example, reduction in morphological attributes of height, coverage, or biomass derived from exposure to HM has been reported in *Arundo donax* (Poaceae), exposed to As, Cd, and Pb [59]; *Zea mays* (Graminaceae) exposed to Cd, Fe, Ni, and Zn [60]; *Prosopis laevigata* (Fabaceae [61]), *Pithecellobium dulce* (Fabaceae [22]), *Vachellia farnesiana* (Fabaceae [21]), and *Wigandia urens* [23] exposed to Cu, Pb, and Zn.

Although the immediate effects of exposure to metals occur at the molecular and cellular levels, they can be extended to higher levels of biological organization: populations, communities, and ecosystems.

In the last decade, one of the emergent effects at the population level that has been evaluated in environmentally exposed populations to HM is shifts in their genetic pool, which were defined by Mussali-Galante and collaborators as permanent biomarkers [12]. Particularly, populations can undergo changes in their diversity and genetic structure [62] in two ways: (1) increased genetic variation as a result of mutations induced by genotoxic agents or (2) decreased genetic variability as a result of bottlenecks or selection [12]. In fact, these changes in the genetic reservoir of populations exposed to HM have been proposed as an indicator of ecosystem health [12]. In general, plant species in which alterations on its genetic diversity have been reported as a consequence of HM exposure include *Taraxacum officinale* (Asteraceae [63]), *Silene paradoxa* (Caryophyllaceae [64]), *Thlaspi caerulescens* (Brassicaceae [65]), *Pinus sylvestris* (Pinaceae [66]), *Thlaspi caerulescens* (Brassicaceae [67]), *Cistus ladanifer* (Cistaceae [68]), *Arabidopsis halleri* (Brassicaceae [69]), and *Prosopis laevigata* (Fabaceae [70]).

HM exposure may have consequences in higher levels of biological organization such as plant communities; however, few studies have examined these effects [71–73]. Shifts in diversity and species richness, changes in dominant species, changes in species composition, and biodiversity loss may be some of the emergent effects. For example, in a study carried out by Martínez-Becerril, it was documented that the vegetal community (trees, shrubs, and herbaceous) associated to mine tailings differed in its species composition compared to reference sites. Likewise, a significant reduction in species richness and diversity was documented [74]. Similarly, a reduction in plant species diversity in grassland contaminated by Cu and Cd [75] and a reduction in vegetal abundance with the increase of HM soil concentration in a metallophyte plant community [76] has been reported using plant communities as the point of interest.

Additionally, HM bioaccumulation in plants may also affect its interactions with other species. It has been suggested that metal accumulation by plants may be a defense strategy to discourage consumption by herbivores [77]. This defensive role of HM has been reported, for example, in the hyperaccumulator *Stanleya pinnata* (Brassicaceae) against black-tailed prairie dog herbivory in seleniferous habitats [78, 79]. Similarly, some species of herbivores have evolved to use metals bioaccumulated in the ingested plant biomass as a defense against subsequent predation [80]. Even, metals can also affect plant-pollinator interactions as reported in *Impatiens capensis* (Balsaminaceae) in sites contaminated with Al and Ni [81].

At the ecosystem level, bioaccumulation within successive trophic levels (biomagnification) has been well documented for some metals. Under this scenario, the plants (primary producers) represent an important step in metal transfer since they constitute the base of the food chain. Therefore, certain metals can be transported from plants to higher levels of the food chain, representing a threat to biodiversity and to ecosystem integrity [82]. For example, HM transfer along the trophic chain has been reported for the Ni hyperaccumulator plant *Alyssum pintodalsilvae* (Brassicaceae) that transfers Ni to grasshoppers (herbivore) and spiders (carnivorous insect), the spiders having higher Ni concentrations [71]. Boyd and Wall found similar results suggesting that Ni could be passed from herbivorous to carnivorous insects [83]. Even, it has been documented that the transfer of HM reaches animals such as small mammals, reporting higher HM levels in carnivorous or omnivorous mammals in comparison to those that feed only by plants [84–86]. This process has also been reported in plants not considered hyperaccumulators. For example, Notten and collaborators report that *Urtica dioica* (Urticaceae) that is distributed in areas with elevated metal concentrations contained only very low metal concentrations [87]. However, the snail *Cepaea nemoralis*, which is the main herbivore feeding on these plants, did contain metal concentrations that were much higher than background values [87, 88]. These studies demonstrate that HM can be transferred among invertebrate species, mobilizing metals from one trophic level to another [84–86]. Finally, these studies evidenced the importance of vegetal species for the evaluation of HM impact on the trophic chain levels, as well as their incorporation and biomagnification patterns.

2.2 Lichens as biosensors

Lichens may be considered as one of the most commonly applied organisms as biosensors [89]. They are symbiotic organisms of fungi and algae and have been widely used in biomonitoring of air pollution [90]. Some of the most commonly used lichen species for toxic metals biomonitoring are *Parmelia sulcata*, *P. caperata*, *Hypogymnia physodes*, and *Xanthoria parietina* [89, 91]. Lacking a protective cuticle and roots, lichens absorb and retain nutrients and trace elements, including HM from dry and wet atmospheric deposition that exceed their physiological requirements. They tolerate these high concentrations by sequestering elements extracellularly as oxalate crystals or lichen acid complexes [92, 93].

A plethora of metal toxicity symptoms, including loss of cell membrane integrity, potassium leakage, disruption of ultrastructure, chlorophyll degradation, and oxidative stress, have been reported in lichens [94–96]. However, their expression depends on the metal and lichen species involved [97, 98]. For example, it is widely known that Cu causes cell membrane damage and adversely influences the photosynthetic apparatus of lichens [99] and also affects fungal and algal ultrastructure [100]. Also GSH (reduced glutathione), the precursor of PCs, is the principal low-molecular thiol and nonenzymatic antioxidant in lichens [101]. It plays a critical role in cellular defense against oxidative damage caused also by HM. For example, in a laboratory study, Pawlik-Skowronska and collaborators found that Cd, Pb, and Zn induced biosynthesis of PCs in the widespread epiphytic lichens *Xanthoria parietina*, *Physconia grisea*, and *Physcia adscendens* [102].

Toxic metal effects reported on lichens include discolored or pinkish necrotic patches and an absence of growth [103]. However, there are tolerant species, some normally associated with HM. For example, *Stereocaulon* species colonized polluted road-site during the period of high Pb emissions [104], and *Vezdaea leprosa* occurs alongside motorway crash barriers in Germany and the UK [105]. So, characteristic lichen assemblages occur on metalliferous soils polluted by industrial emissions and on abandoned mine wastes [105–107] or lichen communities growing on trees [108]. Hence, bark and soil HM contents play a major influence in determining the composition of epiphytic lichen floras [108–110]. Several studies have evidenced the metal influences on epiphytic lichen abundance, cover, richness, and species diversity. For example, in some studies in coniferous forests, correlations between epiphytic lichen abundance and Mn supply were detected [98, 111]. Specifically, a decreasing cover value of the foliose lichen *Hypogymnia physodes* with increasing Mn concentrations in bark or stemflow were repeatedly found in stands of *Picea abies*. In another example, the lichen community associated to *Vachellia farnesiana*, *Prosopis laevigata*, and *Pithecellobium dulce* in exposed sites to HM (Pb, Cu, and Zn) showed higher richness and species diversity values as compared with a reference site [108]. In conclusion, lichens have proved to be very effective organisms as biosensors to detect HM in the environment.

2.3 Transgenic plants as biosensors

The alternative use of transgenic plants in horticulture, forestry, and construction seems to be more appealing for the public. In this respect, design and production of transgenic plants for environmental biomonitoring and cleaning up polluted areas can be action for more favorable public perception of genetically modified organisms [13].

Recently, substantial progress in generation and exploitation of transgenic plants as biomonitors has been made [112, 113]. One of the important advantages of transgenic biosensors is the ability to customize the assay in accordance with monitoring needs. This not only makes transgenic biosensors more sensitive to a particular pollutant but also allows for easy scoring.

Classically, a major approach to addressing these issues has been based on selective breeding or genetic engineering of plants in order to increase their baseline hardiness and/or ability to efficiently utilize resources [114]. Concurrent with these approaches have been efforts to develop and apply technologies toward monitoring and understanding the physiological responses of plants to stress [115–117]. This second, more recent approach is based on leveraging the finely tuned and highly sensitive mechanism plants which have developed to sense, to respond, and to adapt to changes in the environment [118].

Appreciation of this internal decision-making process in plants has led to the development of methods to monitor relevant natural physical phenomena, such as changes in the chlorophyll fluorescence spectra [115, 116]. Another route has been the direct engineering of plants to act as “vital reporters” both of their own health and of internal decision-making processes (so-called biosensors). By coupling knowledge of the genetic cascade stress responses with reporter proteins [e.g., beta-glucuronidase (GUS), luciferase (LUC), or fluorescent proteins (FP)], it is possible to visualize genetic events linked to/associated with stress responses [115, 119]. Indeed, prior research has demonstrated that both endogenous and synthetic promoters can be used as “biomarkers” for a variety of stress conditions, with the appropriate choice of promoter depending on a number of factors, including ease of interpretation, signal-to-noise ratio, and the timeliness of data acquisition [115, 120]. Thus, biomarkers have the potential of informing the researcher, in real time, of the magnitude of a wide variety of physiological events. In particular, the use of FPs has distinct advantages; namely, FP outputs are observable using widely available equipment (e.g., a fluorescent microscope) and require no exogenous additives [118]. With the increasing availability of portable meters for measuring fluoresce [121], it is now feasible to transit this technology from the lab to the greenhouses and the fields.

3. The use of small mammals as biosensors of heavy metal pollution

3.1 Small mammals as biosensors

Small mammals (SM) are frequently used to monitor environmental contamination with HM such as Pb, Cd, Cr, Zn, Al, silver (Ag), As, Co, Cu, Fe, Mn, magnesium (Mg), nickel (Ni), Hg, selenium (Se), strontium (Sr), and Mo. These animals have been used mainly because they are found in the intermediate positions of trophic chains and they are small, have diverse diets, are relatively easy to capture, and have wide geographic distribution (which allows to compare between populations of contaminated and non-contaminated sites). The liver, kidneys, bone, muscle, brain, testicles, teeth, and blood are the main target organs for HM. Conducting studies with SM is important because they allow to make inferences about the bioavailability and bioaccumulation of HM, the biotransformation mechanisms of HM among different species, and the sources of exposure associated with the diet; they also allow to determine which species are susceptible to HM, which is an important step for the evaluation of the biomagnification of HM. Most of the monitoring studies of HM use SM belonging to two orders: Soricomorpha (shrews and moles) and Rodentia (squirrels, rats, mice, voles). The present chapter will focus on two species of the order Rodentia that belong to the families Muridae and Cricetidae. The life history characteristics of both species are described below.

Family Muridae: *Apodemus sylvaticus*; common names include long-tailed field mouse, small wood mouse, and wood mouse. Its conservation status is a minor concern [122]. It has 32 subspecies. Its geographical distribution includes Europe (with the exception of Finland and northern Russia) and some regions of North Africa. It is found at altitudes up to 3300 m.a.s.l. and has been recorded in a variety of seminatural habitats that include forests, moors, steppes, arid Mediterranean scrub, and sand dunes. It is also found in artificial habitats such as suburban and urban parks, gardens, vacant lots, pastures, crops, fields, and forest plantations. It is an omnivorous species that feeds at ground level; its diet includes plants/seeds (70–80%) and invertebrates (20%). It eats tree seeds, fleshy fruits, mushrooms, flowers, and aerial parts of plants. It also consumes fern leaves (*Culcita macrocarpa*) and oak acorns (*Quercus*) [123]. It has also been reported to eat worms, which could be an important source of HM for this species [124]. It is predated by snakes (*Hemorrhois hippocrepis*), eagle-owls (*Bubo bubo*), barn owls (*Tyto alba*), and foxes (*Vulpes vulpes*), among others. It is nocturnal and lives in galleries dug at shallow depths, in crevices, or tree holes. The home range of males is larger than that of females, and it becomes larger for males during the reproductive period. The home range of males can be up to 1.44 ha and 0.49 ha for females [123]. However, some studies estimate that the home range of the males of this species can be up to 2500 m^2 and that its activity range is 56.4 m [125]. The males are polygamous, and, during the breeding season, they travel long distances in search of reproductive partners. There are reports of attacks against intruders and subordinate males, which are thus expelled from the territory and which are displaced from the territories. Unlike males, females have exclusive territories. Their fertility rate is 1–7 l/year, each with an average of 5 pups. The maximum recorded life span in the wild is 12 months [123].

Family Cricetidae: taxonomic name, *Myodes glareolus*; its synonym is *Clethrionomys glareolus*. The common name is bank vole. Its conservation status is a minor concern [126]. A total of 30 subspecies have been reported. It has a wide geographical distribution that includes the British islands, Europe, and Russia. To the north, it can be found beyond the Arctic Circle; to the south, they are found in northern Turkey and Kazakhstan. This species is not found in southern Iberia and the Mediterranean islands. It inhabits altitudes of 2400 m, including open forests, bushes, and hedges [126]. Bank voles are mainly herbivorous, consuming fleshy fruits, seeds, tender leaves, mushrooms, moss, flowers, and roots. They gnaw the bark of young trees and feed on the cambium, but they can also consume earthworms. It is predated by raptors such as the tawny owl (*Strix aluco*), the barn owl (*Tyto alba*), as well as small and medium carnivores [127]. Its home range is estimated to be up to 1000 m^2, and its activity ranges up to 35.7 m [125]. During breeding seasons, the males cover large areas that include the territories of several females. The females have exclusive territories. The mating system is polygynous. The females have 3 or 4 l/reproductive period, with an average of 4 or 5 pups/l. Gestation lasts between 18 and 22 days; the lactation period lasts approximately 18 days. The average life span is between 12 and 13 months, but under extreme conditions they can live for 3 months [127].

3.2 *Apodemus sylvaticus* as a biosensor

There are studies that show that *Apodemus sylvaticus* populations inhabiting contaminated areas bioaccumulate metalloids and HM. For example, Erry and collaborators quantified the concentration of As in *A. sylvaticus* and *C. glareolus* in five sites contaminated with As. These authors found that the organisms of both species accumulate similar concentrations of As in contaminated sites. The concentration of As in the liver and kidneys of the animals inhabiting the contaminated sites was higher than those of animals inhabiting the control site. The concentration of As in those organs was associated with the concentration of As in the stomach contents. Thus, the authors suggest that the animals were exposed to As through the diet and that the two species of mice bioaccumulate As in various organs [128]. Sánchez-Chardi and collaborators compared a population *A. sylvaticus* population inhabiting a control (non-contaminated) site with a population inhabiting a site contaminated by leachates containing potentially toxic elements. They found that the mice inhabiting the leachate site bioaccumulated Cd, Fe, Zn, Cu, Mn, Mo, and Cr, compared with the animals inhabiting the control site. The mice in the leachate site also showed low weight index and a high relative weight of the kidney, as well as high plasma values of glutamic pyruvic transaminase (GPT), an indicator of liver damage. They also showed greater genotoxicity than the animals of the control site. The authors suggest that the morphological and physiological changes observed in the population of *A. sylvaticus* inhabiting the leachate site indicate that this species is more sensitive than *Crocidura russula*, the other studied species inhabiting the site, and that the leachates affected the health of *A. sylvaticus* [129].

The comparison between *A. sylvaticus* and species of the order Soricomorpha, particularly shrews, showed that *A. sylvaticus* is more sensitive to renal toxicity caused by exposure to HM than *C. russula*. Nevertheless, shrews can bioaccumulate more HM. Sánchez-Chardi and collaborators compared populations of *A. sylvaticus* and *C. russula* inhabiting a non-contaminated site (control) with populations of the same species inhabiting a site contaminated by leachates containing potentially toxic elements. In both species inhabiting the contaminated the site, the histological analysis of the liver showed signs of necrosis and apoptosis, inflammation of preneoplastic nodules, and vacuolization. The kidneys were altered mainly in *A. sylvaticus* (necrosis and tubular inflammation), which suggested that this species is more sensitive to renal toxicity than *C. russula* [130]. However, some authors mention that shrews can bioaccumulate more metals and metalloids than *A. sylvaticus*. Mertens and collaborators found that, in contaminated sites (a dredged material deposit), the shrew *Sorex araneus* bioaccumulates more Cd than *A. sylvaticus* and *C. glareolus*. This could be explained by the dietary habits of the studied species, since the diet of *Sorex araneus* consists of invertebrates, including insects and molluscs, while *A. sylvaticus* and *C. glareolus* are mainly herbivorous [131].

Studies by Wijnhoven and collaborators on floodplain species (*A. sylvaticus*, *C. glareolus*, *C. russula*, *M. agrestis*, *M. arvalis*, *M. minutus*, *S. araneus*) found that two species of shrew had higher concentrations of HM compared to the other species; the highest concentrations were found in the shrew *S. araneus*, which has insectivorous and carnivorous habits. Only Cu concentrations were higher in *C. glareolus* than in *A. sylvaticus* and *M. agrestis*. The differences in the concentrations of HM may be due to variations in exposure time (age of the individual), the heterogeneity of the concentrations of HM in soil, the movement of the animals to the other sites, and their feeding patterns. The accumulation of HM in the studied species could also be a risk factor for their predators, potentially altering the structure of their communities and the dynamics of the ecosystem [86].

Cooke and collaborators studied three mammalian species, *A. sylvaticus*, *M. agrestis*, and *S. araneus*, associated with a site contaminated with Pb, Cd, and F. The total accumulation levels of these three compounds in the studied species had the following order: *S. araneus* > *M. agrestis* > *A. sylvaticus*. The stomach contents of *S. araneus* showed that it had the highest intake of Pb, F, and Cd [132]. The differences in bioaccumulation are due to differences in daily intake, in the efficiency of digestion and assimilation, and to other physiological, biochemical, and behavioral factors. Similarly, Drouhot and collaborators found that *Crocidura russula* accumulated more As than *A. sylvaticus*, *Mus spretus*, and *Microtus arvalis*. They also mention that the differences in the accumulation of As between species and within the same species are due to variations in diet, foraging behavior, differences in metabolism, amount of ingested soil, and mobility of the organisms [133].

Some authors have used *A. sylvaticus* in distance gradient studies of contaminated areas. Scheirs and collaborators studied the concentration of metals (Cd, Co, Cr, Cu, Fe, Mn, Pb, and Zn) in soil and the genotoxicity found in *A. sylvaticus* along a distance gradient. The authors reported that the concentration of HM and the genetic damage found in *A. sylvaticus* was higher near the most contaminated areas [134]. Rogival and collaborators studied the accumulation of As, Cd, Cu, and Pb and Zn in *A. sylvaticus* mice inhabiting five sites along a distance gradient, in the soil of the sites, and in the mice's diet (acorns and two species of earthworms: *Dendrodrilus rubidus* and *Lumbricus rubellus*). They observed a gradient in the exposure to metals, beginning on the foundry (most contaminated site), in all the studied elements (soil, diet, and rodent), but not for the essential metals analyzed (Cu and Zn). The concentrations of As, Cd, and Pb in acorns were higher in the sites closest to the foundry. In earthworms, the concentrations of the five metals were higher near the foundry. The transfer of metals occurred mainly from the diet to the mice in the case of Pb and Cd [124]. Another study conducted by Tête and collaborators found that the concentrations of Pb in the liver and kidneys of *A. sylvaticus* followed a distance gradient from the contamination source (foundry). In contrast, the concentrations of Cd in the liver and kidneys of mice varied along the contamination gradient, forming a bell curve. Unlike the results of bioaccumulation, renal alterations (necrosis, lymphocyte infiltration) did not show an increase associated with a distance gradient. The results showed that *A. sylvaticus* is chronically exposed to Pb and Cd and that there is kidney damage present in the species [135].

3.3 *Myodes glareolus* as biosensor

Myodes glareolus, or *Clethrionomys glareolus*, has been used mainly in studies of bioaccumulation of metals and metalloids. Wijnhoven and collaborators analyzed several species of small mammals living in a contaminated floodplain. They found that, in almost 40% of the population of *C. glareolus*, the concentration of Cd exceeded the lowest level at which adverse effects are produced. The other two species, *Microtus agrestis* and *M. arvalis*, showed less ecotoxicological effects [136]. Topashka-Ancheva and collaborators evaluated other small mammals: *A. flavicollis*, *M. macedonicus*, *C. glareolus*, *P. subterraneus*, *M. arvalis*, *M. rossiaemeridionales*, and *C. nivalis*. They found that *C. glareolus* had a higher concentration of Cu and Cd in the body compared to the other species. The concentrations of Cu, Zn, Pb, and Cd in *C. glareolus* were significantly higher than in *A. flavicollis* in both the whole body and in the liver (except for Pb in the liver, which was higher in *A. flavicollis*). The authors suggest that the differences between species are due to the position of each species in the trophic chain, their diet, and lifestyle [137]. Damek-Poprawa and Sawicka-Kapusta compared the populations of a control site and two contaminated sites close to a steel and zinc foundry. No damage was found in *C. glareolus* inhabiting the control site, but there were histopathological changes in the kidneys and liver of the rodents inhabiting the contaminated sites. The concentration of Pb and Cd in liver, kidney, and femur tissues was higher in the rodents living in contaminated areas [138]. Erry and collaborators studied populations of *A. sylvaticus* and *C. glareolus* in a site contaminated with As and in a control site. Many species of rodents living in the contaminated site accumulated more As in the spleen, lung, muscle, and femur than those living in the control site. The concentrations of As in the liver, femur, and hair were higher in *A. sylvaticus* than in *C. glareolus* in both the contaminated and the control sites. The authors mention that these results could be due to the high water exchange and urinary excretion of *C. glareolus* compared to *A. sylvaticus*, which could make *C. glareolus* susceptible to renal toxicity [139].

As shown in the studies on *A. sylvaticus* and *C. glareolus*, the differences between both species are a function of diet, metabolism, mobility, and lifestyle; thus, the monitoring of environmental contamination with metals and metalloids should use small mammals belonging to different taxa in order to determine the real impact of HM on organisms and on trophic chains.

4. The use of bacteria as biosensors of heavy metal pollution

4.1 Bacteria as biosensors

Microorganisms are primary producers in many environmental ecosystems and play an essential role in the nutrient cycle, and they are very abundant and ubiquitous. The microbes proliferate rapidly, are easily detectable and easy to sample, and respond quickly to environmental changes, like temperature, pH, or the presence of contaminants including heavy metals. These characteristics make microorganisms good candidates as pollution biosensors [140]. In this sense, bioluminescent bacteria such as *Aliivibrio fischeri* and *Photobacterium phosphoreum* have been used to monitor water and soil contaminated with HM [141, 142]. This bioassay is carried out using the natural bioluminescence emitted by these bacteria and is based on the decrease of this fluorescence when the bacteria grow in samples of water or soil contaminated with different heavy metals such as Zn, Cu, Cd, Hg, and Cr, among others [141–144]. In the case of *A. fischeri*, the test has been developed commercially and is distributed under the name of Microtox®. However, this method is sensitive to different pollutants such as antibiotics, pesticides, toxins, and organic compounds, which makes it a non-specific method for the detection of heavy metals [145]. Another bacterium proposed as a bioindicator is *Vogesella indigofera*; under normal conditions this bacterium develops a blue color due to the indigoidine production; when it grows in the presence of Cr6, the bacteria decreases the pigment production, and this decrease is dependent on the concentration of Cr^6; at 150 μg/ml the bacteria are entirely white and rough [146]. *Serratia marcescens* is a Gram-negative bacterium that produces a red pigment known as prodigiosin; when the bacterium grows in sub-inhibitory concentrations of Cd, Cr, and Pb, the pigment production decreases drastically, so the authors propose it as a bioindicator of heavy metal contamination [147].

The presence of heavy metals in the environment exerts an intense selection pressure on the organisms that live there; an increase in its concentration and the high rate of horizontal gene transfer can select heavy metal-resistant microorganisms. Therefore, the resistance and detoxification genes have been used as biomarkers for the study of contaminated environments using molecular techniques such as quantitative PCR and real-time quantitative reverse transcription PCR. Within these genes are those involved in the resistance to As, *ACR3*(1) (arsenite efflux pump), *aioA* (arsenite oxidase), *arsB* (arsenical efflux pump), *arsC* (arsenate reductase), and *arsM* (arsenic methyltransferase) [148–151]; those that confer resistance to Cu, *copA* (Copper-exporting P-type ATPase), and *cusA* (copper export system) [152, 153]; for Cd, Zn, and Co resistance, *czcA* (Cd/Zn/Co efflux pump) [154]; for Hg, *hgcA* (mercury methylating protein) and *merA* (mercuric reductase) [155]; the *mr* [140] that encodes to metallothionein a cysteine-rich and heavy metal-binding protein [156]; and *sodA* [140] which codes for a superoxide dismutase, involved in the protection of toxicity against heavy metals [157]. Another technique to measure the presence and abundance of genes involved in resistance to heavy metals is through the use of genetic microarrays such as the GeoChip, commercially available [158]. With the use of this microarray, it was possible to correlate the presence of *arsC*, *copA*, *cueO* (multicopper oxidase), *merB* (alkylmercury lyase), *metC* (cystathionine beta-lyase), *tehB* (tellurite methyltransferase), and *terC* (tellurium resistance protein) genes in sediments and waters contaminated with Cd, Cr, Cu, Hg, and S [158, 159].

High concentrations of heavy metals affect microbial populations and therefore their processes. Thus, the evaluation of microbial processes represents good biomarkers of exposure in different environments. Within the parameters most used are the monitoring of enzymatic activities of the carbon and nitrogen cycle, soil respiration, microbial mass, and the ecosystem biodiversity [160, 161]. Microbial biodiversity is drastically affected by contamination with heavy metals. In general, it is observed that a higher concentration of heavy metals decreases bacterial species. However, with the massive sequencing of DNA, some bacterial groups that could serve as biosensors of contamination were identified, for example, the study carried out by Schneider and collaborators finds that the bacterial groups γ-*Proteobacteria*, *Verrucomicrobia*, and Chlamydiae showed a consistent response to Pb content across contrasting ecosystems. The phyla *Chlamydiae* and γ-*Proteobacteria* were more abundant, while *Verrucomicrobia* were less abundant at high contamination level. So, they conclude that such groups and ratios thereof can be considered as relevant bioindicators of Pb contamination [162]. In soils contaminated with Cu, it was observed that at increased concentrations, bacterial richness was negatively impacted and enhanced relative abundance of *Nitrospira* and *Acidobacteria* members and a lower representation of *Verrucomicrobia*, *Proteobacteria*, and *Actinobacteria*, suggesting a promising role as bioindicators of copper contamination in soils [163].

Acknowledgements

We thank the Consejo Nacional de Ciencia y Tecnología (CONACyT) for the scholarships to IHP.

Conflict of interest

The authors declare that there is no conflict of interest.

Author details

Efraín Tovar-Sánchez[1], Ramón Suarez-Rodríguez[2], Augusto Ramírez-Trujillo[2], Leticia Valencia-Cuevas[1], Isela Hernández-Plata[3] and Patricia Mussali-Galante[3]*

1 Laboratorio de Marcadores Moleculares, Centro de Investigación en Biodiversidad y Conservación, Autonomous University of Morelos State, Cuernavaca, Morelos, Mexico

2 Laboratorio de Fisiología Molecular de Plantas, Centro de Investigación en Biotecnología, Autonomous University of Morelos State, Cuernavaca, Morelos, Mexico

3 Laboratorio de Investigaciones Ambientales, Centro de Investigación en Biotecnología, Autonomous University of Morelos State, Cuernavaca, Morelos, Mexico

*Address all correspondence to: patricia.mussali@uaemmx

References

[1] Sanitá di Toppi L, Gabbrielli R. Response to cadmium in higher plants. Environmental and Experimental Botany. 1999;**41**:105-130. DOI: 10.1016/S0098-8472(98)00058-6

[2] Srivastava NK, Majumder CB. Novel biofiltration methods for the treatment of heavy metals from industrial wastewater. Journal of Hazardous Materials. 2008;**151**:1-8. DOI: 10.1016/j.jhazmat.2007.09.101

[3] Oehme I, Wolfbeis S. Optical sensors for determination of heavy metals ions. Mikrochimica Acta. 1997;**126**(3–4):177-192. DOI: 10.1007/BF01242319

[4] Agency for Toxic Substances and Disease Registry (ATSDR). CERCLA Priority List of Hazardous Substances. Atlanta, GA: Department of Health and Human Services; 2007. Available from: http://www.atsdr.cdc.gov/cercla/07list.html

[5] Kopittke PM, Blamey FPC, Asher CJ, Menzies NW. Trace metal phytotoxicity in solution culture: A review. Journal of Experimental Botany. 2010;**61**:945-954. DOI: 10.1093/jxb/erp385

[6] Patra M, Bhowmik N, Bandopadhyay B, Sharma A. Comparison of mercury, lead and arsenic with respect to genotoxic effects on plant systems and the development of genetic tolerance. Environmental and Experimental Botany. 2004;**52**:199-223. DOI: 10.1016/j.envexpbot.2004.02.009

[7] Lane TW, Saito MA, George GN, Pickering IJ, Prince RC, Morel FMM. A cadmium enzyme from a marine diatom. Nature. 2005;**435**:42. DOI: 10.1038/435042a

[8] Hansch R, Mendel RR. Physiological functions of mineral micronutrients (Cu, Zn, Mn, Fe, Ni, Mo, B, Cl). Current Opinion in Plant Biology. 2009;**12**:259-266. DOI: 10.1016/j.pbi.2009.05.006

[9] Horst WJ, Wang Y, Eticha D. The role of the root apoplast in aluminum-induced inhibition of root elongation and in aluminum resistance of plants: A review. Annals of Botany. 2010;**106**:185-197. DOI: 10.1093/aob/mcq053

[10] Benavides MP, GAllego SM, Tomaro ML. Cadmium toxicity in plants. Brazilian Journal of Plant Physiology. 2005;**17**:21-34. DOI: 10.1590/S1677-04202005000100003

[11] Johnson A, Singhal N, Hashmatt M. Metal-plant interactions: Toxicity and tolerance. In: Khan M, Zaidi A, Goel R, Musarrat J, editors. Biomanagement of Metal-Contaminated Soils, Environmental Pollution. Vol. 20. Dordrecht: Springer; 2011. pp. 29-63. DOI: 10.1007/978-94-007-1914-9_2

[12] Mussali-Galante P, Tovar-Sánchez E, Valverde M, Rojas E. Biomarkers of exposure for assessing environmental metal pollution: From molecules to ecosystems. Revista Internacional de Contaminación Ambiental. 2013;**29**:117-140

[13] Kobalchuk I, Kovalchuk O. Transgenic plants as sensors of environmental pollution genotoxicity. Sensors. 2008;**8**:1539-1558. DOI: 10.3390/s8031539

[14] Stankovic S, Kalaba P, Stankovic AR. Biota as toxic metal indicators. Environmental Chemistry Letters. 2014;**12**:63-84. DOI: 10.1007/s10311-013-0430-6

[15] Nagajyoti PC, Lee KD, Sreekanth TVM. Heavy metals, occurrence and toxicity for plants: A review. Environmental Chemical Letters. 2010;**8**:199-216

[16] Vajpoyee P, Dhawan A, Shanker R. Evaluation of the alkaline comet assay conducted with the wetlands plant *Bacopa monnieri* L. as a model for ecogenotoxicity assessment. Environmental and Molecular Mutagenesis. 2006;**47**:483-489. DOI: 10.1002/em.20217

[17] Gichner T, Patková Z, Száková J, Demnerová K. Toxicity and DNA damage in tobacco and potato plants growing on soil polluted with heavy metals. Ecotoxicologyand Environmental Safety. 2006;**65**: 420-426. DOI: 10.1016/j.ecoenv.2005. 08.006

[18] Maldonado M, Rodríguez C, López A, Wrobel K, Wrobel K. Global DNA methylation in earthworms: A candidate biomarker of epigenetic risks related to the presence of metals/metalloids in terrestrial environments. Environmental Pollution. 2011;**159**: 2387-2392. DOI: 10.1016/j.envpol.2011. 06.041

[19] İşeri IÖD, Körpe DA, Yurtcu E, Sahin FI, Haberal M. Copper-induced oxidative damage, antioxidant response and genotoxicity in *Lycopersicum esculentum* mill. And *Cucumis sativu* L. Plant Cell Reports. 2011;**30**:1713-1721. DOI: 10.1007/s00299-011-1079-x

[20] Murillo HAI. Detección de daño genotóxico en *Prosopis laevigata* de los jales de la Sierra de Huautla, Morelos, México provocado por metales pesados [thesis]. Mexico: Universidad Nacional Autónoma de México; 2015

[21] Santoyo-Martínez M. Bioacumulación, daño genotóxico y cambios en la estructura y morfología foliar de *Vachelia farnesiana* en los jales de Huautla, Morelos [thesis]. Mexico: Universidad Autónoma del Estado de Morelos; 2016

[22] Castañeda-Bautista JA. Estudio ecotoxicológico de los jales mineros de Huautla, Morelos: El caso de *Pithecellobium dulce* (Roxb.) Beth. (Fabaceae) [thesis]. Mexico: Universidad Autónoma del Estado de Morelos; 2016

[23] Cobarrubias-Escamilla DL. Evaluación del daño genotóxico y cambios morfológicos de *Wigandia urens* expuesta a metales pesados en presas de jales abandonados [thesis]. Mexico: Universidad Autónoma del Estado de Morelos; 2017

[24] Ferrer A. Intoxicación por metales. Anales Sis San Navarra. 2003;**26**:141-153

[25] Manara A. Plant responses to heavy metal toxicity. In: Furini A, editor. Plant and Heavy Metals. Netherlands: Springer; 2012. p. 27-44. DOI: 10.1007/ 978-94-007-4441-7_2

[26] Romero-Puertas MC, Corpas FJ, Rodríguez-Serrano M, Gómez M, del Río LA, Sandalio LM. Differential expression and regulation of antioxidative enzymes by cadmium in pea plants. Journal of Plant Physiology. 2007;**164**:1346-1357. DOI: 10.1016/j. jplph.2006.06.018

[27] Dat JF, Vandenabeele S, Vranova E, Van Montagu M, Inze D, Van Breusegem F. Dual action of the active oxygen species during plant stress responses. Cellular and Molecular Life Sciences. 2000;**57**:779-795. DOI: 10.1007/s000180050041

[28] Maksymiec W, Krupa Z. The effects of short-term exposition to Cd, excess Cu ions and jasmonate on oxidative stress appearing in *Arabidopsis thaliana*. Environmental and Experimental Botany. 2006;**57**:187-194. DOI: 10.1016/ j.envexpbot.2005.05.006

[29] Cho U-H, Park J-O. Mercury-induced oxidative stress in tomato seedlings. Plant Science. 2000;**156**:1-9. DOI: 10.1016/S0168-9452(00)00227-2

[30] Yadav K. Heavy metals toxicity in plants: an overview on the role of glutathione and phytochelatins in heavy metal stress tolerance of plants. South African Journal of Botany. 2010;**76**: 167-179. DOI: 10.1016/j.sajb.2009. 10.007

[31] Cobbett C, Goldsbrough P. Phytochelatins and metallothioneins: Roles in heavy metal detoxification and homeostasis. Annual Review in Plant Biology. 2002;**53**:159-182

[32] Hirata K, Tsuji N, Miyamoto K. Biosynthetic regulation of phytochelatins, heavy metal-binding peptides. Journal of Bioscience and Bioengineering. 2005;**100**:593-599. DOI: 10.1263/jbb.100.593

[33] Cobbett CS. Phytochelatins and their roles in heavy metal detoxification. Current Opinion in Plant Biology. 2000; **3**:211-216. DOI: 10.1104/pp.123.3.825

[34] Nocito FF, Pirovano L, Cocucci M, Sacchi GA. Cadmium-induced sulfate uptake in maize roots. Plant Physiology. 2002;**129**:1872-1879. DOI: 10.1104/ pp.002659

[35] Clemens S. Molecular mechanisms of metal tolerance and homeostasis. Planta. 2001;**212**:475-486. DOI: 10.1007/ s004250000458

[36] Inohue M. Phytochelatines. Brazilian Journal of plant physiology. 2005;**17**:65-78. DOI: 10.1007/ s004250000458

[37] Ben Ammar W, Nouairi I, Tray B, Zarrouk M, Jemal F, Ghorbal MH. Effets du cadmium sur l'accumulation ionique et les teneurs en lipides dans les feuilles de tomate (*Lycopersicon esculentum*). Société de Biologie. 2005;**199**:157-163. DOI: 10.1051/jbio:2005016

[38] Drazić G, Mihailović N, Lojić M. Cadmium accumulation in *Medicago sativa* seedlings treated with salicylic acid. Biologia Plantarum. 2006;**50**: 239-244. DOI: 10.1007/s10535-006- 0013-5

[39] Nouairi I, Ben Ammar W, Ben Youssef N, Ben Miled Daoud D, Ghorbal MH, Zarrouk M. Comparative study of cadmium effects on membrane lipid composition of *Brassica juncea* and *Brassica napus* leaves. Plant Science. 2005;**170**:511-519. DOI: 10.1016/j. plantsci.2005.10.003

[40] Pietrini F, Iannelli MA, Pasqualini S, Massacci A. Interaction of cadmium with glutathione and photosynthesis in developing leaves and chloroplasts of *Phragmites australis*. Plant Physiology. 2003;**133**:829-837. DOI: 10.1104/ pp.103.026518

[41] Gouia H, Suzuki A, Brultfert J, Ghorbal MH. Effects of cadmium on the co-ordination of nitrogen and carbon metabolism in bean seedlings. Journal of Plant Physiology. 2003;**160**:367-376. DOI: 10.1078/0176-1617-00785

[42] Agrawal V, Sharma K. Phytotoxic effects of Cu, Zn, Cd and Pb on in vitro regeneration and concomitant protein changes in *Holarrhena antidysenterica*. Biologia Plantarum. 2006;**50**:307-310. DOI: 10.1007/s10535-006-0027-z

[43] Scebba F, Arduini I, Ercoli L, Sebastiani L. Cadmium effects on growth and antioxidant enzymes activities in *Miscanthus sinensis*. Biologia Plantarum. 2006;**50**:688-692. DOI: 10.1007/s10535-006-0107-0

[44] Mohanpuria P, Rana NK, Yadav SK. Cadmium induced oxidative stress influence on glutathione metabolic genes of *Camellia sinensis* (L.) O. Kuntze. Environmental Toxicology. 2007;**22**:368-374. DOI: 10.1002/ tox.20273

[45] Salt DE, Thurman DA, Tomsett AB, Sewell AK. Copper phytochelatins of *Mimulus guttatus*. Procedings of the

Royal Society B. 1989;**236**:79-89. DOI: 10.1098/rspb.1989.0013

[46] Haag-Kerwer A, Schafer HJ, Heiss S, Walter C, Rausch T. Cadmium exposure in *Brassica juncea* causes a decline in transpiration rate and leaf expansion without effect on photosynthesis. Journal of Experimental Botany. 1999;**50**:1827-1835. DOI: 10.1093/jxb/50.341.1827

[47] Schmöger ME, Oven M, Grill E. Detoxification of arsenic by phytochelatins in plants. Plant Physiology. 2000;**122**:793-801. DOI: 10.1104/pp.122.3.793

[48] Ramos J, Naya L, Gay M, Abian J, Becana M. Functional characterization of an unusual phytochelatin synthase, LjPCS3, of *Lotus japonicus*. Plant Physiology. 2008;**148**:536-545. DOI: 10.1104/pp.108.121715

[49] Costa PM, Caeiro S, Diniz MS, Lobo J, Martins M, Ferreira AM, et al. Biochemical endpoints on juvenile *Solea senegalensis* exposed to estuarine sediments: The effect of contaminant mixtures on metallothionein and CYP1A induction. Ecotoxicology. 2009;**18**: 988-1000. DOI: 10.1007/s10646-009-0373-7

[50] Hassinen VH, Tervahauta AI, Schat H, Kärenlampi SO. Plant metallothioneins—Metal chelators with ROS scavenging activity? Plant Biology. 2011;**13**:225-232. DOI: 10.1111/j.1438-8677.2010.00398.x

[51] Zhou J, Goldsbrough PB. Functional homologs of fungal metallothionein genes from Arabidopsis. The Plant Cell. 1994;**6**:875-884. DOI: 10.1105/tpc.6.6.875

[52] Domènech J, Mir G, Huguet G, Capdevila M, Molinas M, Atri-an S. Plant metallothionein domains: Functional insightinto physiological metal binding and protein folding. Biochimie. 2006;**88**:583-593. DOI: 10.1016/j.biochi.2005.11.002

[53] Freisinger E. Metallothioneins in plants. In: Sigel A, Sigel H, Sigel RKO, editors. Metal Ions in LifeSciences. Cambridge, UK. 2009;**5**:107-153. DOI: 10.1039/9781847558992-FP007

[54] Pál M, Holrváth E, Janda T, Páldi E, Szalai G. Physiological changes and defense mechanisms induced by cadmium stress in maize. Journal of Plant Nutrition and Soil Science. 2006; **169**:239-256. DOI: 10.1002/jpln.200520573

[55] Prasad M, Strzalka K. Heavy Metal Stress in Plants. Berlin: Springer; 1999. 138 p. DOI: 10.1007/978-3-662-07745-0_6

[56] Lerda D. The effect of lead on *Allium cepa* L. Mutation Research Letters. 1992;**281**:89-92. DOI: 10.1016/0165-7992(92)90041-F

[57] Eun S, Shik-Youn H, Lee Y. Lead disturbs microtubule organization in the root meristem of *Zea mays*. Physiologia Plantarum. 2000;**110**:357-365. DOI: 10.1111/j.1399-3054.2000.1100310.x

[58] Furini A, editor. Plants and Heavy Metals, Springer Briefs in Biometals. The Netherlands: Springer; 2012. p. 83

[59] Guo Z, Miao X. Growth changes and tissues anatomical characteristics of giant reed (*Arundo donax* L.) in soil contaminated with arsenic, cadmium and lead. Journal of Central South University of Technology. 2010;**17**: 770-777. DOI: 10.1007/s11771-010-0555-8

[60] Tovar-Sánchez E, Cervantes-Ramírez T, Castañeda-Bautista J, Gómez-Arroyo S, Ortiz-Hernández L, Sánchez-Salinas E, et al. Response of *Zea mays* to multimetal contaminated soils: A multibiomarker approach.

Ecotoxicology. 2018;**27**:1-17. DOI: 10.1007/s10646-018-1974-9

[61] Hernández-Lorenzo B. Análisis de la anatomía y morfología de *Prosopis laevigata*, por acumulación de metales pesados en la sierra de Huautla, Morelos [thesis]. Mexico: Universidad Autónoma del Estado de Morelos; 2014

[62] Bickham J, Sandhu S, Herbert P, Chikhi L, Athwal R. Effects of chemical contaminants on genetic diversity in natural populations: Implications for biomonitoring and ecotoxicology. Mutation Research. 2000;**463**:33-51. DOI: 10.1016/S1383-5742(00)00004-1

[63] Keane B, Collier M, Rogstad S. Pollution and genetic structure of North American populations of the common dandelion (*Taraxacum officinale*). Environmental Monitoring an Assessment. 2005;**105**:341-357. DOI: 10.1007/s10661-005-4333-2

[64] Mengoni A, Barabesi C, Gonnelli C, Galardi F, Gabbrielli R, Bazzicalupo M. Genetic diversity of heavy metal-tolerant populations in *Silene paradoxa L.* (Caryophyllaceae): A chloroplast microsatellite analysis. Molecular Ecology. 2001;**10**:1909-1916. DOI: 10.1046/j.0962-1083.2001.01336.x

[65] Basic N, Besnard G. Gene polymorphisms for elucidating the genetic structure of the heavy-metal hyperaccumulating trait in *Thlaspi caerulescens* and their cross-genera amplification in Brassicaceae. Journal of Plant Research. 2006;**119**:479-487. DOI: 10.1007/s10265-006-0011-x

[66] Prus-Glowacki CE, Wojnicka-Pòltorak A, Kozacki L, Fagiewicz K. Effects of heavy metal pollution on genetic variation and cytological disturbances in *Pinus sylvestris L.* population. Journals of Applied Genetics. 2006;**47**:99-108. DOI: 10.1007/BF03194607

[67] Jiménez-Ambriz G, Petit C, Bourrié I, Olivieri I, Dubois S, Ronce O. Life history variation in the heavy metal tolerant plant *Thlaspi caerulescens* growing in a network of contaminated and non-contaminated sites in southern France: Role of gene flow, selection and phenotypic plasticity. New Phytologist. 2007;**173**:199-215. DOI: 10.1111/j.1469-8137.2006.01923.x

[68] Quintela-Sabarís C, Glusppe V, Castro-Fernández D, Fraga M. Chloroplast microsatellites reveal that metallicolous populations of the Mediterranean shrub *Cistus ladanifer* L have multiple origins. Plant and Soil. 2010;**334**:161-174. DOI: 10.1007/s11104-010-0368-4

[69] Godé C, Decombeix I, Kostecka A, Wasowicz P, Pauwels M, Courseaux A, et al. Nuclear microsatellite loci for *Arabidopsis halleri* (Brassicaceae), a model species to study plant adaptation to heavy metals. American Journal of Botany. 2012;**99**:e49-e52. DOI: 10.3732/ajb.1100320

[70] Fuentes-Reza A. Bioacumulación y análisis de la estructura y diversidad genética de *Prosopis laevigata* por exposición a metales en Huautla, Morelos [thesis]. Mexico: Universidad Autónoma del Estado de Morelos; 2017

[71] Peterson LR, Trivett V, Baker AJM, Aguiar C, Pollard AJ. Spread of metals through an invertebrate food chain as influenced by a plant that hyperaccumulates nickel. Chemoecology. 2003;**13**:103-108. DOI: 10.1007/s00049-003-0234-4

[72] Boyd RS, Wall MA, Jaffré T. Nickel levels in arthropods associated with Ni hyperaccumulator plants from an ultramafic site in New Caledonia. Insect Science. 2006;**13**:271-277. DOI: 10.1111/j.1744-7917.2006.00094.x

[73] Bourioug M, Gimbert F, Alaoui-Sehmer L, Benbrahim M, Aleya L,

Alaoui-Sossé B. Sewage sludge applicationin a plantation: Effects on trace metal transfer in soil-plant-snail continuum. Science of the Total Environment. 2015;**502**:309-314. DOI: 10.1016/j.scitotenv.2014.09.022

[74] Martínez-Becerril C. Efecto de los metales pesados en jales mineros sobre la comunidad vegetal de la microcuenca de Huautla y el daño genotóxico en su herbívoro dominante *Peromyscus levipes* (Rodenta: Muridae) [thesis]. Mexico: Universidad Autónoma del Estado de Querétaro; 2009

[75] Hunter B, Johnson M, Thompson D. Ecotoxicology of copper and cadmium in acontaminated grassland ecosystem. Jurnal of Applied Ecology. 1989;**26**: 89-99. DOI: 10.2307/2403653

[76] Becerril JM, Barrutia O, Plazaola JG, Hernández A, Olano JM, Garbisu C. Ecosistemas especies nativas de suelos contaminados por metales: Aspectos ecofisiológicos y su uso en fitorremediaciøn. Ecosistemas: Revista científica y técnica de ecología y medio ambiente; 2007;**16**:50-55

[77] Boyd RS. Plant defense using toxic inorganic ions: Conceptual models of the defensive enhancement and joint effects hypotheses. Plant Science. 2012;**195**:88-95. DOI: 10.1016/j.plantsci.2012.06.012

[78] Quinn CF, Freeman JL, Galeas ML, Klamper EM, Pilon-Smits EAH. The role of selenium in protecting plants against prairie dog herbivory: Implications for the evolution of selenium hyperaccumulation. Oecologia. 2008; **155**:267-275. DOI: 10.1007/s00442-007-0907-8

[79] Freeman JL, Quinn CF, Lindblom SD, Klamper EM, Pilon-Smits EAH. Selenium protects the hyperaccumulator *Stanleya pinnata* against black-tailed prairie dog herbivory in native seleniferous habitats. American Journal of Botany. 2009;**96**:1075-1085. DOI: 10.3732/ajb.0800287

[80] Boyd RS. High-nickel insects and nickel hyperaccumulator plants: A review. Insect Science. 2009;**16**:19-31. DOI: 10.1111/j.1744-7917.2009.00250.x

[81] Meindl GA, Ashman TL. The effects of aluminum and nickel in nectar on the foraging behavior of bumblebees. Environmental Pollution. 2013;**177**:78-81. DOI: 10.1016/j.envpol.2013.02.017

[82] Peralta-Videa JR, Lopez ML, Narayan M, Saupe GJ. The biochemistry of environmental heavy metal uptake by plants: Implications for the food chain. The International Journal of Biochemistry & Cell Biology. 2009;**41**: 1665-1677. DOI: 10.1016/j.biocel.2009.03.005

[83] Boyd RS, Wall MA. Responses of generalist predators fed high-Ni *Melanotrichus boydi* (Heteroptera: Miridae): Elemental defense against the third trophic level. The American Midland Naturalist. 2001;**146**:186-198. DOI: 10.1674/0003

[84] Hamers T, van den Berg JHJ, van Gestel CAM, van Schooten FJ, Murk AJ. Risk assessment of metals and organic pollutants for herbivorous and carnivorous small mammal food chains in a polluted floodplain (Biesbosch, The Netherlands). Environmental Pollution. 2006;**144**:581-595. DOI: 10.1016/j.envpol.2006.01.020

[85] Veltman K, Huijbregts MAJ, Hamers T, Wijnhoven S, Hendriks AJ. Cadmium accumulation in herbivorous and carnivorous small mammals: Meta-analysis of field data and validation of the bioaccumulation model optimal modeling for ecotoxicological applications. Environmental Toxicology and Chemestry. 2007;**26**:1488-1496. DOI: 10.1897/06-518R.1

[86] Wijnhoven S, Leuven RSE, Van der Velde G, Jungheim G, Koelemij EI, de

Vries FT, et al. Heavy-metal concentrations in small mammals from a diffusely polluted floodplain: Importance of species-and location-specific characteristics. Archives of Environmental Contamination and Toxicology. 2007;**52**:603-613. DOI: 10.1007/s00244-006-0124-1

[87] Notten MJM, Oosthoek AJP, Rozema J, Aerts R. Heavy metal concentrations in a soil-plant-snail food chain along a terrestrial soil pollution gradient. Environmental Pollution. 2005;**138**:178-190. DOI: 10.1016/j.envpol.2005.01.011

[88] Notten MJM, Oosthoek AJP, Rozema J, Aerts R. The landsnail *Cepaea nemoralis* regulates internal Cd levels when fed on Cd-enriched stinging nettle (*Urtica dioica*) leaves at low, field-relevant concentrations. Environmental Pollution. 2006;**139**:296-305. DOI: 10.1016/j.envpol.2005.05.007

[89] Paoli L, Corsini A, Bigagli V, Vannini J, Bruscoli C, Loppi S. Long-term biological monitoring of environmental quality around a solid waste landfill assessed with lichens. Environmental Pollution. 2012;**161**:70-75. DOI: 10.1016/j.envpol.2011.09.028

[90] Purvis OW, Pawlik-Skowrońska B. Lichens and metals. In: Avery S, Stratford M, van West P, editors. Stress in Yeasts and Filamentous Fungi. Amsterdam: Elsevier; 2008. pp. 175-200. DOI: 10.1016/S0275-0287(08)80054-9

[91] Nimis PL, Lazzarin G, Lazzarin A, Skert N. Biomonitoring of trace elements with lichens in Veneto (NE Italy). Science of the Total Environment. 2000;**255**:97-111. DOI: 10.1016/S0048-9697(00)00454-X

[92] Giordani P, Modenesi P, Tretiach M. Determinant factors for the formation of the calcium oxalate minerals, weddellite and whewellite, on the surface of foliose lichens. The Lichenologist. 2003;**35**:255-270. DOI: 10.1016/S0024-2829(03)00028-8

[93] Pawlik-Skowrońska B, Purvis OW, Pirszel J, Skowroński T. Cellular mechanisms of Cu tolerance in the epilithic lichen *Lecanora polytropa* growing at a copper mine. The Lichenologist. 2006;**38**:267-275. DOI: 10.1017/S0024282906005330

[94] Branquinho C. Lichens. In: Prasad MNV, editor. Metals in the Environment: Analysis by Biodiversity. New York: Marcel Dekker; 2001. pp. 117-115

[95] Paul A, Hauck M, Langenfeld-Heyser M. Ultrastructural changes in soredia of the epiphytic lichen Hypogymnia physodes cultivated with manganese. Environmental and Experimental Botany. 2004;**52**:139-147. DOI: 10.1016/j.envexpbot.2003.01.001

[96] Sanita di Toppi L, Marabottini R, Vattuone Z, Musetti R, Favali MA, Sorgona A, et al. Cell wall immobilisation and antioxidant status of *Xanthoria parietina* thalli exposed to cadmium. Functional Plant Biology. 2005;**32**:611-618. DOI: 10.1071/FP04237

[97] Branquinho C, Brown DH, Catarino F. The cellular location of Cu in lichens and its effects on membrane integrity and chlorophyll fluorescence. Environmental and Experimental Botany. 1997;**38**:165-179. DOI: 10.1016/S0098-8472(97)00015-4

[98] Hauck M, Paul A. Manganese as a site factor for epiphytic lichens. The Lichenologist. 2005;**37**:409-423. DOI: 10.1017/S0024282905014933

[99] Garty J, Tamir O, Levin T, Lehr H. The impact of UV-B and Sulphur- or copper-containing solutions in acidic conditions on chlorophyll fluorescence in selected *Ramalina* species. Environmental Pollution. 2007;**145**:266-273. DOI: 10.1016/j.envpol.2006.03.022

[100] Tarhanen S. Ultrastructural responses of the lichen *Bryoria fuscescens* to simulated acid rain and heavy metal deposition. Annals of Botany. 1998;**82**: 735-746. DOI: 10.1006/anbo.1998.0734

[101] Kranner I, Cram WJ, Zorn M, Wornik S, Yoshimura I, Stabentheiner E, et al. Antioxidant and photoprotection in a lichen as compared with its isolated symbiont partners. Proceedings of the National Academy of Sciences of the United States of America. 2005;**102**:3141-3146. DOI: 10.1073/pnas.0407716102

[102] Pawlik-Skowrońska B, di Toppi LS, Favali MA, Fossati F, Pirszel J, Skowronski T. Lichens respond to heavy metals by phytochelatin synthesis. New Phytologist. 2002;**156**:95-102. DOI: 10.1046/j.1469-8137.2002.00498.x

[103] Seaward MRD. Some observations on heavy metal toxicity and tolerance in lichens. The Lichenologist. 1974;**6**:158-164. DOI: 10.1017/S0024282974000260

[104] Ernst G. *Vezdaea leprosa*—Spezialist am strassenrand. Herz. 1995; **11**:175-188

[105] Gilbert OL. Lichens. London: Harper Collins; 2000

[106] Aptroot A, van den Boom PPG. *Pyrenocollema chlorococcum*, a new species with a chlorococcoid photobiont from zinc-contaminated soils and wood. Cryptogamie Bryologie Lichenologie. 1998;**19**:193-196

[107] Cuny D, Denayer FO, de Foucault B, Schumacker R, Colein P, Van Haluwyn C. Patterns of metal soil contamination and changes in terrestrial cryptogamic communities. Environmental Pollution. 2004;**129**: 289-297. DOI: 10.1016/j.envpol.2003.10.009

[108] Eligio-González S. Comunidades liquénicas asociadas a *Pithecellobium dulce*, *Acacia farnesiana* y *Prosopis laevigata* en jales mineros y sitios testigos en Huautla, Morelos [tesis]. Cuernavaca, Morelos: Universidad Autónoma del Estado de Morelos; 2016

[109] Gauslaa Y, Holien H. Acidity of boreal *Picea abies*-canopy lichens and their substratum, modified by local soils and airborne acidic depositions. Flora. 1998;**193**:249-257. DOI: 10.1016/S0367-2530(17)30845-9

[110] Díaz-Escandón D. Líquenes cortícolas como indicadores ambientales en los alrededores de mina de azufre El Vinagre (Cauca) [thesis]. Santiago de Cali, Colombia: Programa Académico de Biología; 2012

[111] Hauck M. Epiphytic lichen diversity and forest dieback: The role of chemical site factors. The Bryologist. 2003;**106**:257-269. DOI: 10.1639/00072745(2003)106[0257:ELDAFD]2.0.CO;2

[112] Kobalchuk I, Kovalchuk O, Arkhipov A, Hohn B. Transgenic plants are sensitive bioindicators of nuclear pollution caused by the Chernobyl accident. Nature Biotechnology. 1998; **16**:1054-1059. DOI: 10.1038/3505

[113] Kovalchuk O, Titov V, Hohn B, Kobalchuk I. A sensitive transgenic plant system to detect toxic inorganic compounds in the environment. Nature Biotechnology. 2001;**19**:568-572. DOI: 10.1038/89327

[114] Varshney RK, Bansal KC, Aggarwal PK, Datta SK, Craufurd PQ. Agricultural biotechnology for crop improvement in a variable climate: Hope or hype? Trends in Plant Science. 2011;**16**: 363-371. DOI: 10.1016/j.tplants.2011.03.004

[115] Liew OW, Chong PCJ, Li B, Asundi AK. Signature optical cues: Emerging technologies for monitoring plant

health. Sensors. 2008;**8**:3205-3239. DOI: 10.3390/s8053205

[116] Chaerle L, Lenk S, Leinonen I, Jones HG, Van Der Straeten D, Buschmann C. Multi-sensor plantimaging: Towards the development of a stress-catalogue. Biotechnology Journal. 2009;**4**:1152-1167. DOI: 10.1002/biot.200800242

[117] Wee CW, Dinney JR. Tools for high-spatial and temporal-resolution analysis of environmental responses in plants. Biotechnology Letters. 2010;**32**: 1361-1371. DOI: 10.1007/s10529-010-0307-8

[118] Hines G, Modavi C, Jiang K, Packard A, Poolla K, Feldman L. Tracking transience: A method for dynamic monitoring of biological events in *Arabidopsis thaliana* biosensors. Plant. 2015;**242**:1251-1261. DOI: 10.1007/s00425-015-2393-2

[119] Meyer AJ, Brach T, Marty L, Kreye S, Rouhier N, Jacquot JP, et al. Redox-sensitive GFP in *Arabidopsis thaliana* is a quantitative biosensor for the redox potential of the cellular glutathione redox buffer. Plant Journal. 2007;**52**(5): 973-986. DOI: 10.1111/j.1365-313X. 2007.03280.x

[120] Antunes MS, Morey KJ, Smith JJ, Albrecht KD, Bowen TA, Zdunek JK, et al. Programmable ligand detection system in plants through a synthetic signal transduction pathway. PLoS One. 2001;**6**(1):e16292. DOI: 10.1371/journal. pone.0016292

[121] Millwood RJ, Halfhill MD, Harkins D, Russotti R, Stewart CN Jr. Instrumentation and methodology for quantifying GFP fluorescence in intact plant organs. BioTechniques. 2003;**34**(3):638-643. DOI: 10.2144/03343pf02

[122] Schlitter D, van der Straeten E, Amori G, Hutterer R, Kryštufek B, Yigit N, Mitsain G. *Apodemus sylvaticus*. The IUCN Red List of Threatened Species. 2016:e.T1904A115059104. DOI: 10.2305/IUCN.UK.2016-3.RLTS. T1904A22423831.en

[123] Sánchez-González B, Navarro-Castilla A, Hernández MC, Barja I. Ratón de campo-*Apodemus sylvaticus*. In: Salvador A, Barja I, editors. Enciclopedia Virtual de los Vertebrados Españoles. Madrid, España: Museo Nacional de Ciencias Naturales; 2016. Available from: http://www.vertebradosibericos.org/

[124] Rogival D, Scheirs J, Blust R. Transfer and accumulation of metals in a soil-diet-wood mouse food chain along a metal pollution gradient. Environmental Pollution. 2007;**145**: 516-528. DOI: 10.1016/j.envpol.2006. 04.019

[125] Wijnhoven S, Van Der Velde G, Leuven RSEW. Smits AJM flooding ecology of voles, mice and shrews: The importance of geomorphological and vegetational heterogeneity in river floodplains. Acta Theriologica. 2005; **50**(4):453-472. DOI: 10.1007/ BF03192639

[126] Hutterer R, Kryštufek B, Yigit N, Mitsain G, Palomo LJ, Henttonen H, et al. Myodes glareolus. The IUCN red list of threatened species 2016:e. T4973A11 5070929. DOI: 10.2305/IUCN. UK.2016-3.RLTS.T4973A22372716.en

[127] Luque-Larena JJ, Gosálbez Noguera J. *Myodes glareolus* (Schreber, 1780). In: Palomo LJ, Gisbert y J, Blanco JC, editors. Atlas y libro rojo de los mamíferos terrestres de España. Dirección General para la Biodiversidad -SECEM-SECEMU, Madrid, España. 2007. 398-400

[128] Erry BV, Macnair MR, Meharg AA, Shore RF. Arsenic contamination in wood mice (*Apodemus sylvaticus*) and bank voles (*Clethrionomys glareolus*) on

abandoned mine sites in Southwest Britain. Environmental Pollution. 2000; **110**(1):179-187. DOI: 10.1016/S0269-7491(99)00270-5

[129] Sánchez-Chardi A, Peñarroja-Matutano C, Ribeiro CA, Nadal J. Bioaccumulation of metals and effects of a landfill in small mammals. Part II. The wood mouse, *Apodemus sylvaticus*. Chemosphere. 2007;**70**(1):101-109. DOI: 10.1016/j.chemosphere.2007.06.047

[130] Sánchez-Chardi A, Peñarroja-Matutano C, Borrás M, Nadal J. Bioaccumulation of metals and effects of a landfill in small mammals part III: Structural alterations. Environmental Research. 2009;**109**(8):960-967. DOI: 10.1016/j.envres.2009.08.004

[131] Mertens J, Luyssaert S, Verbeeren S, Vervaeke P, Lust N. Cd and Zn concentrations in small mammals and willow leaves on disposal facilities for dredged material. Environmetal Pollution. 2001;**115**:17-22. DOI: 10.1016/S0269-7491(01)00096-3

[132] Cooke JA, Andrews SM, Johnson MS. Lead, zinc, cadmium and fluoride in small mammals from contaminated grassland established on fluorspar tailings. Water, Air, and Soil Pollution. 1990;**5**:43-54. DOI: 10.1007/BF00211502

[133] Drouhot S, Raoul F, Crini N, Tougard C, Prudent AS, Druart C, Rieffel D, Lambert JC, Tête N, Giraudoux P, Scheifler R. Responses of wild small mammals to arsenic pollution at a partially remediated mining site in southern France. Science of the Total Environment. 2014;470-471:1012-1022. DOI: 10.1016/j.scitotenv.2013.10.053.14

[134] Scheirs J, De Coen A, Covaci A, Beernaert J, Kayawe VM, Caturla M, et al. Getoxicity in wood mice (*Apodemus sylvaticus*) along a pollution gradient: Exposure-, age-, and gender-related effects. Environmental Toxicology and Chemestry. 2006;**25**(8): 2154-2162. DOI: 10.1897/05-419R.1

[135] Tête N, Durfort M, Rieffel D, Scheifler R, Sánchez-Chardi A. Histopathology related to cadmium and lead bioaccumulation in chronically exposed wood mice, *Apodemus sylvaticus*, around a former smelter. Science Total Environment. 2014;**481**: 167-177. DOI: 10.1016/j.scitotenv.2014.02.029

[136] Wijnhoven S, Leuven RS, van der Velde G, Eijsackers HJ. Toxicological risks for small mammals in a diffusely and moderately polluted floodplain. Science Total Environment. 2008; **406**(3):401-406. DOI: 10.1016/j.scitotenv.2008.05.059

[137] Topashka-Ancheva M, Metcheva R, Teodorova S. A comparative analysis of the heavy metal loading of small mammals in different regions of Bulgaria I: Monitoring points and bioaccumulation features. Ecotoxicology Environmental Safety. 2003;**54**(2):176-187. DOI: 10.1016/S0147-6513(02)00052-0

[138] Damek-Poprawa M, Sawicka-Kapusta K. Histopathological changes in the liver, kidneys, and testes of bank voles environmentally exposed to heavy metal emissions from the steelworks and zinc smelter in Poland. Environmental Research. 2004;**96**(1): 72-78. DOI: 10.1016/j.envres.2004.02.003

[139] Erry BV, Macnair MR, Meharg AA, Shore RF. The distribution of arsenic in the body tissues of wood mice and bank voles. Archives of Environmental Contamination and Toxicology. 2005; **49**(4):569-576. DOI: 10.1007/s00244-004-0229-3

[140] Bouskill NJ, Barnhart EP, Galloway TS, Handy RD, Ford TE. Quantification of changing *Pseudomonas aeruginosa sodA*, *htpX* and *mt* gene abundance in

response to trace metal toxicity: A potential in situ biomarker of environmental health. FEMS Microbiology Ecology. 2007;**60**:276-286. DOI: 10.1111/j.1574-6941.2007.00296.x

[141] An J, Jeong S, Moon HS, Jho EH, Nam K. Prediction of Cd and Pb toxicity to *Vibrio fischeri* using biotic ligand-based models in soil. Journal of Hazardous Materials. 2012;**203–204**: 69-76. DOI: 10.1016/j.jhazmat. 2011.11.085

[142] Qu R, Wang X, Liu Z, Yan Z, Wang Z. Development of a model to predict the effect of water chemistry on the acute toxicity of cadmium to *Photobacterium phosphoreum*. Journal of Hazardous Materials. 2013;**262**:288-296. DOI: 10.1016/j.jhazmat.2013.08.039

[143] Wang X, Qu R, Wei Z, Yang X, Wang Z. Effect of water quality on mercury toxicity to *Photobacterium phosphoreum*: Model development and its application in natural waters. Ecotoxicology and Environment Safety. 2014;**104**:231-238. DOI: 10.1016/j. ecoenv.2014.03.029

[144] Tsiridis V, Petala M, Samaras P, Hadjispyrou S, Sakellaropoulos G, Kungolos A. Interactive toxic effects of heavy metals and humic acids on *Vibrio fischeri*. Ecotoxicology Environment Safety. 2006;**63**:158-167. DOI: 10.1016/j. ecoenv.2005.04.005

[145] Abbas M, Adil M, Ehtisham-Ul-Haque S, Munir B, Yameen M, Ghaffar A, et al. *Vibrio fischeri* bioluminescence inhibition assay for ecotoxicity assessment: A review. Science of the Total Environment. 2018;**626**: 1295-1309. DOI: 10.1016/j. scitotenv.2018.01.066

[146] Gu JD, Cheung K. Phenotypic expression of *Vogesella indigofera* upon exposure to hexavalent chromium, Cr^{6+}. World Journal of Microbiology and Biotechnology. 2001;**17**:475-480. DOI: 10.1023/A:1011917409139

[147] Cristani M, Naccari C, Nostro A, Pizzimenti A, Trombetta D, Pizzimenti F. Possible use of *Serratia marcescens* in toxic metal biosorption (removal). Environmental Science and Pollution Research. 2012;**19**:161-168. DOI: 10.1007/s11356-011-0539-8

[148] Poirel J, Joulian C, Leyval C, Billard P. Arsenite-induced changes in abundance and expression of arsenite transporter and arsenite oxidase genes of a soil microbial community. Research Microbiology. 2013;**164**:457-465. DOI: 10.1016/j.resmic.2013.01.012

[149] Jiang Z, Li P, Jiang D, Wu G, Dong H, Wang Y, et al. Diversity and abundance of the arsenite oxidase gene *aioA* in geothermal areas of Tengchong, Yunnan. China. Extremophiles. 2014;**18**:161-170. DOI: 10.1007/s00792-013-0608-7

[150] Zhai W, Wong MT, Luo F, Hashmi MZ, Liu X, Edwards EA, et al. Arsenic methylation and its relationship to abundance and diversity of *arsM* genes in composting manure. Scientific Reports. 2017;**7**:42198. DOI: 10.1038/ srep42198

[151] Zhang SY, Zhao FJ, Sun GX, Su JQ, Yang XR, Li H, et al. Diversity and abundance of arsenic biotransformation genes in paddy soils from southern China. Environmental Science and Technology. 2015;**49**:4138-4146. DOI: 10.1021/acs.est.5b00028

[152] Besaury L, Bodilis J, Delgas F, Andrade S, De la Iglesia R, Ouddane B, et al. Abundance and diversity of copper resistance genes *cusA* and *copA* in microbial communities in relation to the impact of copper on Chilean marine sediments. Marine Pollution Bulletin. 2013;**67**:16-25. DOI: 10.1016/j. marpolbul.2012.12.007

[153] Li XF, Yin HB, Su JQ. An attempt to quantify Cu-resistant microorganisms in a paddy soil from Jiaxing, China.

Pedosphere. 2012;**22**:201-205. DOI: 10.1016/S1002-0160(12)60006-X

[154] Roosa S, Wattiez R, Prygiel E, Lesven L, Billon G, Gillan DC. Bacterial metal resistance genes and metal bioavailability in contaminated sediments. Environmental Pollution. 2014;**189**:143-151. DOI: 10.1016/j.envpol.2014.02.031

[155] Liu YR, Yu RQ, Zheng YM, He JZ. Mint: Analysis of the microbial community structure by monitoring an Hg methylation gene (hgcA) in paddy soils along an Hg gradient. Appl Environ Microbiol. 2014;**80**:2874-2879. DOI: 10.1128/AEM.04225-13

[156] Geslin C, Llanos J, Prieur D, Jeanthon C. The manganese and iron superoxide dismutases protect *Escherichia coli* from heavy metal toxicity. Research in Microbiology. 2001;**152**:901-905. DOI: 10.1016/S0923-2508(01)01273-6

[157] Ruttkay-Nedecky B, Nejdl L, Gumulec J, Zitka O, Masarik M, Eckschlager T, et al. The role of metallothionein in oxidative stress. International Journal of Molecular Science. 2013;**14**:6044-6066. DOI: 10.3390/ijms14036044

[158] Saxena G, Marzinelli EM, Naing NN, He Z, Liang Y, Tom L, et al. Ecogenomics reveals metals and land-use pressures on microbial communities in the waterways of a megacity. Environmental Science and Technology. 2015;**49**:1462-1471. DOI: 10.1021/es504531s

[159] Jie S, Li M, Gan M, Zhu J, Yin H, Liu X. Microbial functional genes enriched in the Xiangjiang River sediments with heavy metal contamination. BMC Microbiology. 2016;**16**:179. DOI: 10.1186/s12866-016-0800-x

[160] Brookes PC. The use of microbial parameters in monitoring soil pollution by heavy metals. Biology and Fertility Soils. 1995;**19**:269-279. DOI: 10.1007/BF00336094

[161] Xia X, Lin S, Zhao J, Zhang W, Lin K, Lu Q, et al. Toxic responses of microorganisms to nickel exposure in farmland soil in the presence of earthworm (*Eisenia fetida*). Chemosphere. 2018;**192**:43-45. DOI: 10.1016/j.chemosphere.2017.10.146

[162] Schneider AR, Gommeaux M, Duclercq J, Fanin N, Conreux A, Alahmad A, et al. Response of bacterial communities to Pb smelter pollution in contrasting soils. Science of the Total Environment. 2017;**605-606**:436-444. DOI: 10.1016/j.scitotenv.2017.06.159

[163] Nunes I, Jacquiod S, Brejnrod A, Holm PE, Johansen A, Brandt KK, Priemé A, Sørensen SJ. Coping with copper: legacy effect of copper on potential activity of soil bacteria following a century of exposure. FEMS Microbiology Ecology. 2016;**92**:fiw175. DOI: 10.1093/femsec/fiw175

Chapter 9

Development of an Ethanol Biosensor Based on Silver Nanoparticles/Polyaniline/Graphite/Epoxy Composite for Friendly Analytical Application

Eliana Mossé Alhadeff, Alfredo Jackson Telles Bosco, Caio Fragale Pastusiak, Thais Anjos Correia and Ninoska Isabel Bojorge Ramirez

Abstract

Electrochemical biosensors for measuring ethanol were developed in this study. Silver nanoparticles were incorporated to increase sensibility. Firstly, the transducer element was prepared with silver nanoparticles/polyaniline/graphite/epoxy composite (AgNPs/PANI/GEC) where the enzymes *alcohol oxidase* (AOD) and *horseradish peroxidase* (HRP) were immobilized by adsorption. The composition of the immobilized solution was indicated by an experimental design (85% of the bi-enzymatic solution, 10% of albumin and 5% of glutaraldehyde). The immobilization method adopted in this study showed to be highly reproducible. The values of variance and standard deviation were low (0.003 and 0.053, respectively—means of three electrodes). The linear range was 0–30 g/L (R^2 = 0.983) and the sensitivity 0.004 mA L/g. A second biosensor was made with a transducer prepared with AgNPs/PANI/GEC. A chitosan film was applied over the adsorbed enzymatic solution to avoid desorption of the immobilized enzymes. The AgNPs/PANI/GEC electrodes were characterized using cycle voltammetry and the composite surface by scanning electron microscopy. The calibration for ethanol samples with this second composite fitted in a range of 0.0–0.35 g/L (R^2 = 0.984). Square rate voltammetry was the electrochemical analytical method used to obtain the amperometric calibration curves.

Keywords: alcohol oxidase, horseradish peroxidase, polyaniline, silver nanoparticles, electrochemical biosensor

1. Introduction

Biosensors with electrochemical transducers are attractive from a commercial point of view due to its low expense and simplicity of construction and because it is easy to employ in a broad range of applications and portable [1, 2]. The voltammetric and amperometric technique analyses are a full implementation to measure low levels of concentration. Additionally, as an electrochemical surface method, it has the advantage of generating a response signal working with small sample volumes, in real time. The working electrode can be prepared by surface immobilization of one or more enzymes, which are involved in recognition of the analyte. The enzymes could be regarded as the critical component of the electrode since they are related to the selectivity of the sensor to catalyze the formation of the electroactive detection product [3]. Additionally, the enzymes are ideal biological recognition elements in the construction of biosensors, because of their high specificity, which enables the development of analytical methods with high accuracy [1]. Biosensor technologies are recognized as an emerging science to produce analytical devices that can help to detect specific compounds in complex mixtures such as liquid waste residues and blood serum [4, 5]. Nowadays, ethanol analysis is fundamental for criminal justice systems, in clinical and toxicological diagnostic analyses such as blood, serum and urine analysis, monitoring medical conditions of HIV patients, as well as public safety issues regarding the pilots and drivers. In the food and beverage industries, the determination of alcohol content is critical for the control of the fermentation process and product quality. Besides these applications, determination of ethanol is also important in agricultural, biofuel and environmental analyses [6–8]. *Alcohol oxidase* (AOD) allows the qualitative and quantitative determination of ethanol or methanol, removal of alcohol or aldehyde and hydrogen peroxide production and removal of oxygen [9]. *Horseradish peroxidase* (HRP) uses the haem group and hydrogen peroxide substrate to oxidize a variety of organic and inorganic compounds [10]. The enzymatic mechanisms of these enzymes together facilitate the detection of AOD substrates by electrochemical and spectroscopic methods such as ethanol. The co-immobilization of the enzymes holds the potential to increase the selectivity and amplify the sensitivity of the biosensor improving the potential for quantitative ethanol detection [11, 12]. Polyaniline (PANI) has attracted the attention of the scientific community in the last two decades. PANI, a family of conductive semi-flexible polymers in a green protonated emeraldine form had high electrical conductivity and low production cost. PANI has been explored for various applications, including those in biosensors due to some useful features such as redox conductivity and polyelectrolyte characteristics, high surface area, chemical characteristics, long-term environmental stability and tunable properties enhancing the electron transfer flux ability and also the reversibility of the electrochemical response signal [12–14]. In the previous work, the electroconductivity of PANI-GEC was reported and verified that PANI is an attractive polymer compound to be applied in sensor interface transducer biosensors. A conductivity of 28 μS/cm for 30 w/w% PANI-GEC was measured and cyclic voltammograms for 10 mM potassium ferricyanide obtained working with a scan rate of 100 mV s^{-1}. The composite can act as an effective mediator in the transference of electrons in redox or enzymatic reactions [14]. With the use of nanomaterials, greater sensitivity and attachment of enzymes are achieved due to their high surface area as well as the physical, chemical and electronic properties. The literature reported that nanomaterial application had attracted much attention regarding the development of high-performance electrochemical biosensors [13]. The preparation of chemically modified electrodes with silver nanoparticles (AgNPs) has been applied to amplify the electrochemical response signal. Since silver is four times cheaper than gold and shows excellent catalytic activity and good electrical/heat conductivity, its application is very favorable in electroanalysis acting as a pre-concentrator of species of interest and/or mediating redox reactions [15]. The literature also reported an increase of 9.71% in the anodic peak current and 32.35% for the cathode peak current in the presence of AgNPs. The authors observed an increase in the reversibility of the voltammetric response signal in the composite based on AgNPs/PANI/GEC and the ratio of the anodic (Ipa) and cathodic (Ipc) peak currents $|Ipa/Ipc| = 1.07$ at 40 mV s^{-1}. Nevertheless, the ratio was of the order of 1.28 for composite without AgNPs [16]. From a point of view of analytical instruments, biosensors are used also for quality control, because they have important technical characteristics, such as low response time, high selectivity, stability under the conditions of the analysis, and reproducibility of the measurements. So, ethanol as a hydrous biofuel (Brazil, 5.3 volume % of water) or in the anhydrous form mixed in blends with gasoline (USA—10%, 15% and 85% of anhydrous ethanol, western Europe—5% of anhydrous ethanol and Brazil—27% of anhydrous ethanol) have being used in regular cars and flex fuel vehicles. Methods for monitoring the percentage of ethanol in the mixtures (product quality control) or in case of spilling of gasohol blends need to be develop. Many efforts for different potentiometric, amperometric, and spectrophotometric biosensors have been developed for ethanol analyses, with microorganisms like *Gluconobacter oxydans, Saccharomyces ellipsoideus,* or enzymes as alcohol dehydrogenase or alcohol oxidase were just reported in the literature [17]. In Brazil, there is a rigorous program for a strict control of the physicochemical characteristics of the gasohol blend and hydrated fuel alcohol to prevent adulteration or environmental contamination. Nowadays, the 4.0 industry claims for more robust and sensitive instruments for long-distance transmission and data transfer systems to an analytical central station in monitoring and process control program. Only a few biosensors are commercially available at present for analysis control and the integration of nanomaterials composites within these enzymatic biosensors brings new strategies for enhancing their analytical performances [16, 17]. The high ethanol solubility turns the assessment and analytical methods limited. Many road or pipeline accidents can spill fuel blends into the environment. Significant environmental impacts related to ethanol spills have been to surface water and fishes were killed several days after as a result of oxygen depletion. Spilled ethanol from the surface through soil to groundwater contamination is also of concern, and anaerobic biodegradation of ethanol in groundwater results in the production of methane [18]. The development of more selective and integrated systems for application in fast and high accuracy analysis needs further innovation and research investments. The enzyme immobilization methods represent an important step for the new technologies applied in bioinstrumentation techniques. The incorporation of silver nanoparticles in two different composites and the electrochemical response signals generated from each biosensor were investigated. An experimental design was used and statistic analysis to define the best condition for low cost enzymatic immobilization method. The AgNPs/PANI/GEC biosensor with AOD and HRP immobilized enzymes was firstly prepared to detect ethanol. A second composite with only AgNPs/GEC was prepared and the immobilized AOD and HRP adsorbed enzymes covered with a chitosan film. Voltammetry/amperometric techniques were applied to characterize the electrochemical transduction systems. Calibration curves were obtained for each composite electrode biosensor in order to evaluate the ethanol analytical ranges and detection levels.

2. Electrode construction and characterization

The composite used in the manufacture of the electrode for the biosensor comprised 40% of PANI, 35% epoxy and 25% enriched with graphite AgNPs as described in [16]. The mixture was inserted into the empty end of the Teflon support (rod of 50 mm length × 7 mm in outer diameter, recessed 3 mm diameter × 3 mm deep). The electrodes containing the composites were maintained in an incubator at 30°C for 24 h, and then, the end polished with 1200 mesh sandpaper. In this work two strategies of enzyme immobilization were studied, one by adsorption on the surface composite and a second way covering the bi-enzymatic solution with a film of chitosan polymer. **Figure 1** shows an illustration of the composite surface with

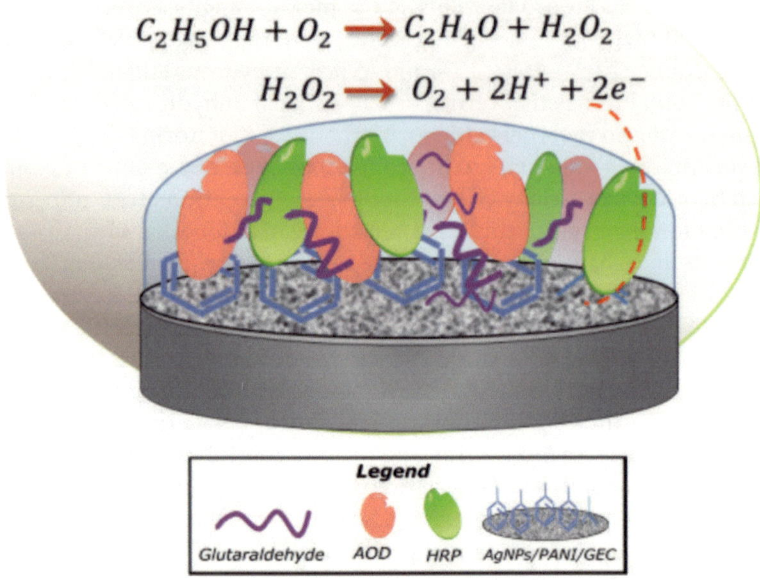

Figure 1.
Schematic diagram of the surface composite with the immobilized AOD and HRP enzymes and the reaction electrode sequence.

enzymes immobilized by adsorption without the chitosan film addition, with the two sequential enzymatic reactions occurring. The ethanol is firstly oxidized to acetaldehyde, and hydrogen peroxide is also produced by the action of the immobilized AOD. After that, the hydrogen peroxide is decomposed in oxygen and water by the immobilized HRP. The oxi-reduction reaction generates electrons by the hydrogen peroxide hydrolysis, and the mechanism could be expressed as follows [19]:

$$\text{HRP (reduced form)} + H_2O_2 \rightarrow \text{HRP (oxidized form)} + H_2O \quad (1)$$

$$\text{HRP (oxidized form)} + 2e^- + 2H^+ \rightarrow \text{HRP (reduced form)} + H_2O \quad (2)$$

$$\text{Net reaction:} H_2O_2 + 2e^- + 2H^+ \rightarrow 2H_2O \quad (3)$$

The chitosan film was prepared by dripping a chitosan solution onto the immobilized enzymes adsorbed on the composite surface. In order to prepare the chitosan solution, 2.0 g was dissolved with 100 ml of acetic acid 0.5% (v/v) and the resulting solution rested without mixing for 2 hours and, in sequence, mixed for 6 h at 1000 rpm. A vacuum filtration pump was used to remove the chitosan particles that did not dissolved completely, and pH was adjusted to 4.9 with acetic acid [20]. The prepared electrodes with chitosan film were stored at room temperature for 6 h and after at 8°C. The reagents used in this research were polyaniline (emeraldine salt), epoxy resin DER 332, glutaraldehyde and enzyme *alcohol oxidase* EC 1.1.3.13 (obtained from *Pichia pastoris*) purchased from Sigma-Aldrich; sodium citrate, 4-aminoantipyrine, hydrogen peroxide, phenol and ethanol 95% P.A. (analytical grade) from Vetec; silver nitrate (Synth); and enzyme *horseradish peroxidase* (EC 1.11.1.7) from Toyobo, Brazil. The lyophilized enzyme was suspended in phosphate buffer pH 7.0, filtered on qualitative-grade filter paper (INLAB) and placed on Spectrum® dialysis membrane (12,000–14,000 kDa) at 8°C immersed in deionized water that was exchanged every 12 hours for 24 h. The purified enzyme was kept in a freezer. Graphite powder (Fluka) was used in the preparation of composites. Sodium phosphate reagent and albumin used in immobilization experiments purchased from Sigma-Aldrich. The composite based in graphite was enriched with a colloidal dispersion of AgNPs synthesized by the Turkevich method [21]. The distribution of AgNPs was characterized by UV–visible spectrophotometer (Shimadzu, model UV—1800 120 V) reading the absorbance at a scanning wavelength between 250 and 700 nm [22, 23]. Atomic force microscopy (AFM) images were taken to identify the AgNPs in the freshly prepared colloidal suspension using an atomic force microscopic Solver NEXT, NT-MDT Integration Solutions for Nanotechnology, SPM Controller (P9 XPM Systems Digital Control Platform), model BL900 [23]. Tyndall effect or the scattering of light by colloidal particles in suspension was observed for recently obtained silver nanoparticles in water using a laser pointer [21]. Scanning electron microscopy (SEM) images were also taken with a scanning electron microscopy—EDS: BAL-TEC SCD 005. Square wave voltammetry was the electrochemical analysis method applied to the generated response signal characterization for different ethanol solution concentrations. The parameters used during the tests with square wave voltammetry were 10 mL of the electrolyte solution, frequency 25 Hz scan rate of 40 mV/s potential between 0.2 and 0.75 V. A potentiostat (Autolab, PGSTAT12 model), reference electrode Ag/AgCl (ALS model RE-012167 1B) and counter electrode platinum wire were used. D-optimal mixture design was used to evaluate the influence of variables on the parameters of the immobilization method based on optimal composition proposed in the literature [24, 25]. The experimental design of the mixture was prepared with the following restrictions: glutaraldehyde volume ranging from 1 to 10%, albumin volume of 0.5–10% and volume of enzyme solution ranging from 80 to 98.5%. The resulting peak current (mA), generated by the formulations tested, was the response signal adopted to evaluate the performance of electrochemical bi-enzymatic biosensors. The mixture response surfaces and principal component analysis calculations were performed using the Statistica 8.0 software. The components used in the manufacture of the lock solution were 2.5% (v/v) glutaraldehyde, albumin 1% (w/v) and an enzymatic solution composed of AOD (286 U) and HRP (2640 U). **Table 1** shows the proposed compositions generated by the mixture experimental design software.

From the proportions suggested by the statistical analysis, 10 μL of immobilization solution was deposited on the electrode surface which remained stored at 4°C for 12 h. Peak currents obtained by square wave voltammetry using parameters were described in Section 2.4. An electrolytic cell containing 10 mL of sodium phosphate buffer (pH 7.0) with the addition of 0.5 mL of 95% ethanol solution was used. Calibration of the AgNPs/PANI/GEC biosensor was performed under optimized conditions at 0.4 V versus AgCl in 0.1 M sodium phosphate buffer solution (pH 7.0).

Experiment	Bi-enzymatic solution (%)	Glutaraldehyde (%)	Albumin (%)
1	80.00	10.0	10.0
2	98.50	01.0	0.50
3	89.00	01.0	10.0
4	89.50	10.0	0.50
5	89.25	5.50	5.25

Table 1.
Percent composition proposed by the design experiments.

10 μL of an immobilization solution, whose experimental design estimates provide the greatest resulting peak current within the planned boundaries, was deposited on the composite electrode surface, which remained stored at 4°C for 12 h. For the construction of the standard curve, aliquots of 95% ethanol PA (concentration: 789 g/L) were added to obtain different concentrations of ethanol. The resulting peak currents were obtained by square wave voltammetry carried out in triplicate. The investigation of repeatability of the biosensor was performed using 9.85 μL of sodium phosphate buffer solution (pH 7.0) with the addition of 0.15 μL of 95% ethanol PA (concentration: 789 g/L) making a concentration of 12.8 g/L of ethanol.

2.1 AgNPs colloidal dispersion characterization

The absorption spectrum of AgNPs colloidal dispersion is shown in **Figure 2(A)**, which exhibits an absorption band at approximately 400 nm, consistent with the results reported in the literature [21, 22]; typical absorption band is in the region of 350–450 nm, confirming that their synthesis was successful. Such bands are unique physical properties of these nanoparticles. When an external magnetic field, such as light, is applied to metal, the conduction electrons move in tandem to provide a distributed load disturbance known as plasma, located close to the metal surface [21]. A relationship between the color of the colloidal silver nanoparticles with the diameter (6–28 nm for yellow color) and the shape was just demonstrated in other researches [23, 24]. **Figure 2(B)** shows the atomic force microscopy measurements of the colloidal AgNPs solution, to show the surface topography of the as-formed silver nanoparticles is homogeneous and uniformly modified on the colloidal solution to obtain a surface-based nanocomposite system like published in previous studies [23–28].

The surface of the graphite/AgNPs composite was prepared by mixing 150 mg of graphite powder with 50 mL of the nanosilver colloidal suspension. A dry composite was obtained after 12 h at 100°C for complete water evaporation [16]. The enriched graphite with the silver nanoparticles, after homogenization, was characterized by MEV images. **Figure 3** shows the rugged surface of the composite used to immobilize the bi-enzymatic solution.

The characterization of the graphite/AgNPs composite surface by SEM generated an irregular and rough surface (image magnification of 2500 times **Figure 3(A)**) that was better seen when amplified at 6.500 times as shown in **Figure 3(B)**. The corresponding selected area energy dispersive spectrometer (EDS) analysis was defined by the yellow circle in **Figure 3(B)**. **Table 2** shows the percentage values (wt/wt%) of the elements in the AgNPs/GEC sample characterized by EDS.

The method used in this work generated a very low level of silver nanoparticles in the composite mixture, but even a few quantities of dispersed silver nanoparticles on graphite power had an important effect on the voltammetric response signal as reported previously in the literature [16] for a AgNPs/PANI/GEC. The authors observed that the AgNPs insertion resulted in an increase of both generated current peaks, anodic and cathodic, for the electrodes prepared with 25% graphite/AgNPs, 40% of PANI and 35% of epoxy resin. The system showed reversibility character, demonstrating that the graphite mixture, PANI, epoxy resin and AgNPs increase the electroconductivity of the electrode.

2.2 Enzyme solution composition statistical analysis

Table 3 shows the averages of the resultant peak currents according to each enzyme immobilization solution.

Figure 4 shows the three-dimensional response surface and the corresponding contour plot obtained by the statistical experimental design. The high, intermediate

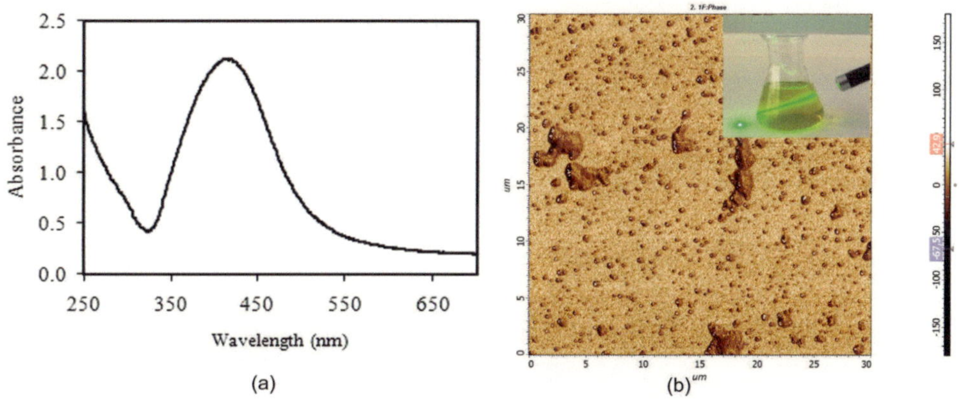

Figure 2.
(A) Absorption spectrum of the AgNPs colloidal dispersion with a maximum at 400 nm. (B) AFM 2D image of silver nanoparticle colloid on the surface composite sensor (scan scale: 30 μm × 30 μm; height scale: 5.0 nm). Top window in (B): Tyndall effect for colloidal AgNPs suspension.

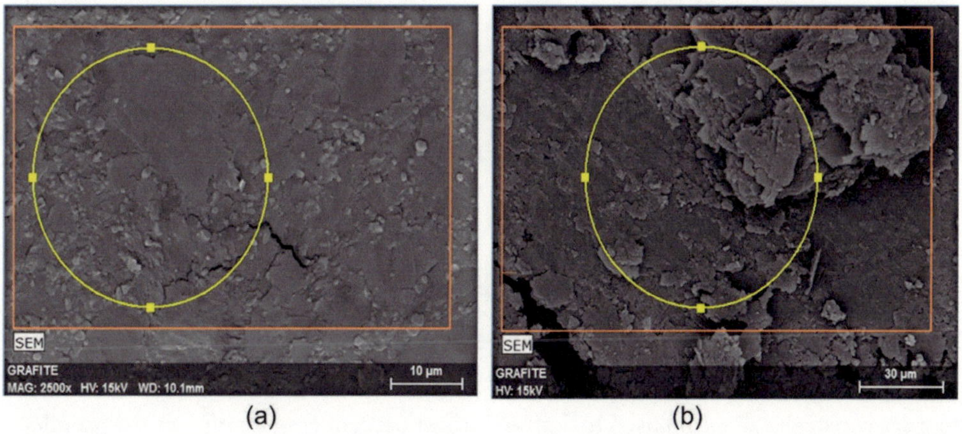

Figure 3.
Surface scanning microscopy photograph of the graphite/AgNPs composite: (A) 2.500× and (B) 6.500×.

Element	Weight (%)
Carbon	83.96
Oxygen	15.62
Silver	0.42
Total	100

Table 2.
Composite composition by EDS analysis.

and low bands of the resulting peak current are produced in the region from green to red depending on the intensity of the generated peak current. For the range of the three pseudo-components in the mixture (enzyme solution, glutaraldehyde, and albumin) investigated in this work, a two-dimensional triangle plot can represent the contours as shown in **Figure 4(A)**. The generated contour lines are based on the constraints of the experimental planning 80–98.5% for the bi-enzymatic solution, 1–10% for glutaraldehyde (2.5%) and 0.5–10% for albumin solution (1% w/v). **Figure 4(B)** shows a

Experiment	Ip_c (mA)	Variance	SD[a]
1	0.015	9.33×10^{-06}	3.06×10^{-03}
2	0.155	3.10×10^{-05}	5.57×10^{-03}
3	0.143	6.16×10^{-04}	2.48×10^{-02}
4	0.181	2.86×10^{-04}	1.69×10^{-02}
5	0.134	4.93×10^{-05}	7.02×10^{-03}

[a]*Standard deviation.*

Table 3.
Amperometric peaks current means.

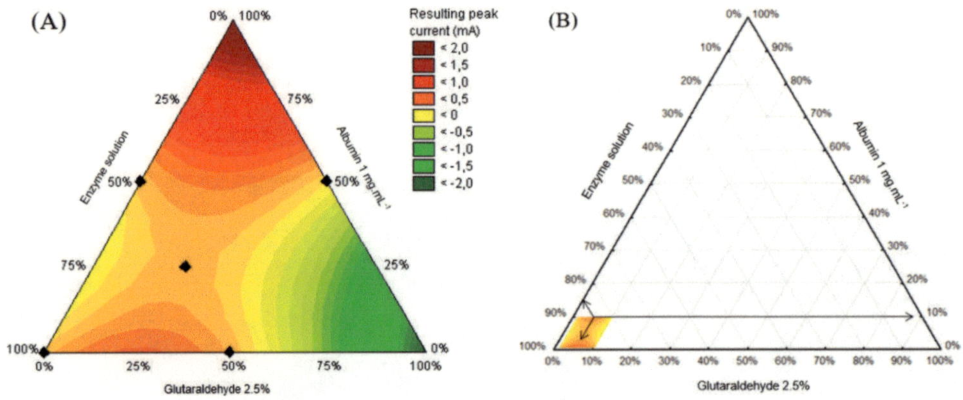

Figure 4.
(A) Level curves for the resulting peak current produced by the electrode with immobilized enzymes according to the pseudo-components of the mixture. (B) Ternary diagram restriction for the resulting peak current produced by the electrode with immobilized enzymes depending on the components, the enzyme solution, albumin and glutaraldehyde.

ternary constraint diagram used to define the best percent composition of the pseudo-component mixture suitable to achieve the highest peak current.

Figure 4(B) shows the planning area is surrounded by a diamond, and the two regions that provide the largest resulting peak currents appear in darker orange. The correct reading of the percentage of each pseudo-component for the mixture is highlighted with arrows, which indicate the direction side of each percentage in the ternary constraint diagram. Two dark orange areas were found in the investigated planning region. The first region indicates a resultant high-peak current supplied by an electrode with immobilized enzymes containing a solution composed of 85% enzyme solution, 10% albumin and 5% glutaraldehyde. The second region indicates a resultant high-peak current delivered by an electrode with immobilized enzyme and volume solution containing 94.5% of the bi-enzymatic solution, 0.5% albumin and 5% glutaraldehyde. The composition of the immobilization solution selected for this study was that of the first region (85% enzyme solution), a lower enzyme content and high-peak current. **Table 4** shows the regression coefficients of pseudo-components used to identify the relevant components and their interactions.

These results indicate that the enzyme solution factor is statistically significant for a confidence interval (CI) of 95% since their level p-value is less than 0.05. Glutaraldehyde, albumin and the interaction of second-order AB (enzyme solution interacting with glutaraldehyde) are marginally significant for the same confidence interval, for $0.05 < p$ level < 0.10. The second-order interaction AC (enzyme

Factor	Coefficient	p (CI: 95%)[a]
(A) Enzymatic solution	0.0001873	0.0000315
(B) Glutaraldehyde 2.5% (%)	−0.0044553	0.0982712
(C) Albumin 1 (%, w/v)	0.0042506	0.0962228
AB	0.0091370	0.0862246
AC	−0.0083963	0.1053623

[a] *CI: confidence interval.*

Table 4.
Regression coefficients of pseudo-components and statistical significance.

solution interacting with albumin) is not statistically significant at the same confidence interval level and $p > 0.10$.

2.3 Calibration curve for the AgNPs/PANI/graphite biosensor

The construction of the calibration curve plotted in **Figure 5(B)** was obtained from the data shown in **Table 5**. These results show the values of peak currents (mA) obtained with the ethanol concentrations utilized in this study (0–30 g/L) are shown in **Figure 5(A)**. An increase of the response signal, proportional to the level of the ethanol concentration in the sample, was observed. The linearity was fitted by Eq. (4):

$$I(mA) = 0.004 Ethanol(g/L) + 0.118 \quad (4)$$

with a coefficient correlation R^2 of 0.983. The sensitivity expressed by the angular coefficient of the linear adjustment is equal to 0.004 (mA L/g). In **Table 5** the variance and standard deviation values are shown for a mean of three new biosensors recently prepared for the measurements of each ethanol concentration solution.

The standard curve shows a positive linearity confirming the oxidation of ethanol by AOD. Ethanol oxidation occurs on the biosensor surface resulting in a stream of electrons at this location. Within the established ethanol concentration range applied in this study, an increase in the ethanol concentration into the electrolytic cell will generate a greater flow of electrons confirming the results reported previously for bi-enzymatic biosensors [29–32].

The limit of detection (LOD) was calculated by Eq. (5) and the limit of quantification (LOD) by Eq. (6) [33]:

$$LOD = \frac{3S_b}{a} \quad (5)$$

$$LOQ = \frac{10\,S_b}{a} \quad (6)$$

in which S_b is the standard deviation of the measurements in white and a the sensitivity of technique. For the AgNPs/PANI/GEC biosensor, the LOD and LOQ calculated values were 3.54 g/L and 11.8 g/L, respectively.

The repeatability was determined by repetitive measures (five) by square wave voltammograms and refers to the agreement between successive measurements of the same sample for each working electrode [33]. The reproducibility was described as the agreement between results (signals) obtained with the same method with three recently prepared biosensors (B1, B2 and B3) [34]. For five sequential analyses, the relative square deviation values (RSD) were 1.54 (B1), 1.01 (B2) and 2.11% (B3)

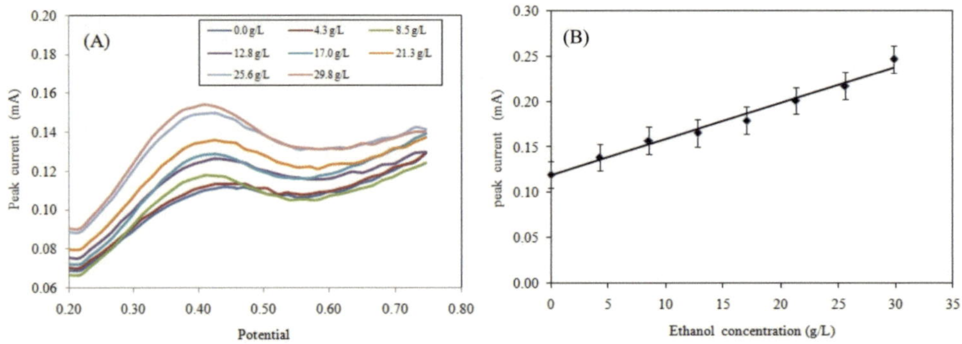

Figure 5.
(A) Peak current generated by square wave voltammetry (mean values of triplicate biosensors measurements). (B) Calibration curve for the bi-enzymatic ethanol biosensor.

Ethanol[a] (g/L)	Current[b] (mA)	Variance	SD[c]
0.0	0.120	2.23×10^{-05}	4.73×10^{-03}
4.3	0.139	6.33×10^{-06}	2.52×10^{-03}
8.5	0.157	4.00×10^{-06}	2.00×10^{-03}
12.8	0.165	4.33×10^{-06}	2.08×10^{-03}
17.0	0.179	7.00×10^{-06}	2.65×10^{-03}
21.3	0.201	3.90×10^{-05}	6.24×10^{-03}
25.6	0.217	6.33×10^{-06}	2.52×10^{-03}
29.8	0.247	2.03×10^{-05}	4.51×10^{-03}

[a]*Diluted ethanol concentration with sodium phosphate buffer pH 7.0.*
[b]*Mean of three measurements.*
[c]*Standard deviation.*

Table 5.
Ethanol biosensor amperometric responses.

for each tested biosensor. The reproducibility values are shown in **Table 6** as the media value from measurements determined with three different biosensors recently built. The variance, standard deviation and RSD values are also shown in **Table 6**.

The results were considered satisfactory as variance and standard deviation values were around 10^{-3} and 10^{-2}, respectively, for the analytical determinations carried out with standard ethanol solution and three freshly prepared working bio-electrodes. To investigate a biosensor with higher sensitivity and accuracy and lower cost production, a composite prepared only with graphite and silver nanoparticles was also studied. The new working electrodes received a film of chitosan covering the immobilized enzymes that were adsorbed on the recently prepared composite. The response time for the square wave analysis was 37 s.

2.4 Biosensor prepared with graphite/AgNPs

2.4.1 Composite electrochemical characterization

Firstly, cyclic voltammetry was run in order to evaluate the reversibility of the electrochemical response signal for the AgNPs/GEC used to build the working electrode. The generated voltammetry curves were investigated for different voltage velocity (10–100 mV/s), and the profiles were shown in **Figure 6(A)** for electrodes

Analyses	Ethanol[a] (g/L)	Variance[b]	SD[c]	RSD[d] (%)
1	15.32	0.003	0.053	0.35
2	15.24	0.006	0.076	0.50
3	15.11	0.003	0.059	0.39
4	14.86	0.015	0.121	0.81
5	14.79	0.017	0.130	0.88

[a]*Media value from measures determined with three different biosensors recently built.*
[b]*Variance values.*
[c]*Standard deviation.*
[d]*Relative standard deviation.*

Table 6.
Variation of ethanol concentration for the reproducibility.

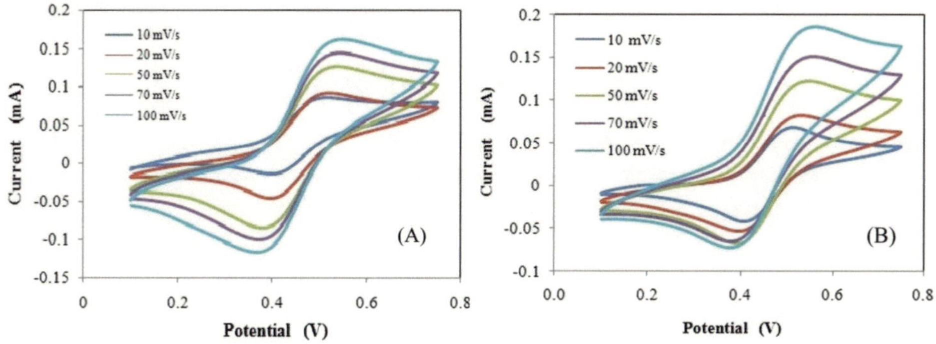

Figure 6.
Cyclic voltammetry for sensors electrodes in $K_3[Fe(CN)_6]$,/KCl pH 7.0 solution. Scanning run of the applied voltage velocity (10, 20, 50, 70, 100 mV/s). (A) Freshly prepared composite electrodes and (B) after 70 days.

built with freshly prepared composite. Electrodes prepared 70 days before were also tested and the cyclic response signals plotted in **Figure 6(B)**.

According to the Randles-Sevcik equation (Eq. (7)) for cycle voltammetry analysis, a linear correlation can exist between cathode or anode peak current, and the velocities square root can be applied during the scanning experiments (**Figure 7**) [35]:

$$I_P = 0.4463nF\sqrt{\frac{nFD}{RT}}AC\sqrt{\upsilon} \quad (7)$$

where I_p is the peak current, n is the number of electrons, F is the Faraday constant, T is the temperature in Kelvin, R is the gas constant, A is the surface area of the working electrode, D is the diffusion coefficient of the electroactive species, C is the bulk concentration of the electroactive species and υ is the scan rate of voltammograms. Thus, the diffusion coefficients for ferrocene and ferricenium at 298 K are calculated from the slope of the plot of I_p versus $\sqrt{\upsilon}$.

The linear correlations are shown in **Table 7** for the electrodes prepared with graphite/epoxy composite (A), with AgNPs/GEC stored for 70 days (B) and with AgNPs/GEC recently prepared (C). The linearity (R^2) for both types of current Ipc and Ipa is close to 1.00 (R^2 = 0.997–0.998) suggesting that the system features a diffusion mass transfer.

The $|Ip_c/Ip_a|$ ratio values are shown in **Table 8**, where (Ip_c) is the reverse current and (Ip_a) the input current for the three electrode conditions studied and calculated for each velocity rate applied during the scanning experiments.

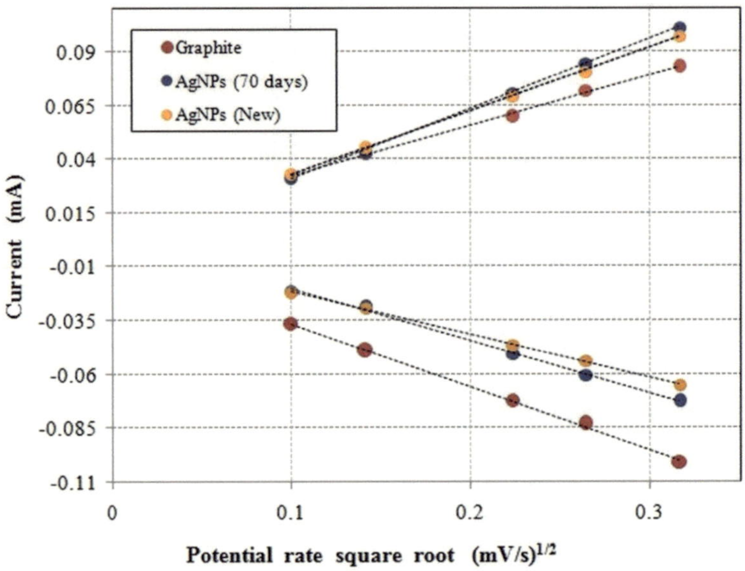

Figure 7.
Linear correlations for cathode and anode peaks current with the velocity square roots ($K_3[Fe(CN)_6]/KCl$ 10 ml solution, pH 7.0).

	Anode	R^2	Cathode	R^2
A	$I = 0.294(\upsilon)^{1/2} + 0.003$	0.999	$I = -0.200(\upsilon)^{1/2} - 0.001$	0.998
B	$I = 0.325(\upsilon)^{1/2} + 0.001$	0.999	$I = -0.243(\upsilon)^{1/2} - 0.004$	0.998
C	$I = 0.235(\upsilon)^{1/2} + 0.008$	0.997	$I = -0.291(\upsilon)^{1/2} - 0.007$	0.998

Table 7.
Linear correlations for current and scanning velocity square rate.

ν (mV/s)	Ip_c/Ip_a		
	A	B	C
10	1.041	1.148	1.155
20	1.141	1.207	1.145
50	1.165	1.131	1.215
70	1.239	1.147	1.152
100	1.234	1.145	1.217

Table 8.
Average values of cathode and anode peak current ratios obtained for three electrodes of the same characteristics and method of preparation showed in item 2.4.1.

The reversibility characteristic of a voltammetric cycle is verified when the |Ipc/Ipa| ratio is equal to 1.0. For the three different composites investigated in this study, the reversibility was observed when low voltammetric rates (mV/s) were applied to the voltammetric cycle analysis. The graphite-epoxy composite electrodes recently built showed a reversible performance (|Ipa/Ipc| = 1.041) for a scanning rate of 10 mV/s. For the AgNPs/GEC electrodes stored during 70 days at 8°C, working with 50 mV/s, a quasi-reversible system was observed (|Ipa/Ipc| = 1.131). The biosensor

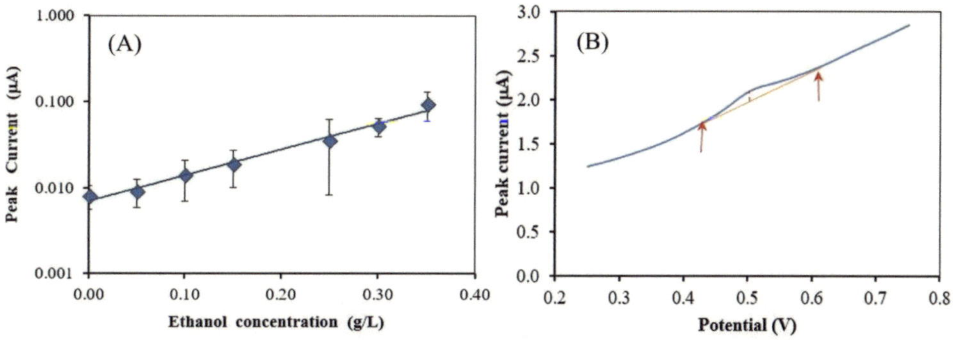

Figure 8.
Ethanol electrochemical calibration (A). Screen printed data for the electrochemical analysis of a square wave voltammetry curve for a sample prepared with 0.35 mL of ethanol and 9.65 mL of phosphate buffer (B).

built with AgNPs/GEC recently prepared also showed a quasi-reversible (|Ipa/Ipc| = 1.145) behaviour for 20 mV/s. These results are similar to those obtained by other authors [1, 3, 9], but with biosensor complex designs, compared to ours.

2.4.2 Calibration curve for the graphite/AgNPs biosensor

Square wave voltammetry was applied to generate the resulting peaks for standard ethanol solutions tested for the new calibration curve. The square wave analyses were conducted with 20 Hz, amplitude of 20 mV and step potential of 1 mV from 250 to 750 mV. The response time chitosan ethanol bi-enzymatic AgNPs/GEC biosensor to 0.35 g/L was 208 s. The working electrodes were prepared (triplicates) and used to analyze ethanol in a range of 0.0–0.35 g/L in phosphate buffer pH 7.0. The intensity of peak current (I) generated for each measurement was determined and media of the resulting values correlated with respective ethanol concentrations. The obtained data were well fitted (R^2 = 0.984) as an exponential function: $I(\mu A) = 0.007e^{6.899[\text{Ethanol, g/L}]}$ shown in **Figure 8(A)**. A curve for an electrochemical analysis sample obtained with the software General Purpose Electrochemical System (GPES) 4.9.005 version, Metrohm, Utrecht, Holland, is showed in **Figure 8(B)**.

The LOD for the AgNPs/GEC biosensor was 3.48×10^{-3} and 0.0116 g/L the value for LOQ. The smaller square wave electrochemical response signals obtained were due to the chitosan applied over the bi-enzymatic solution adsorbed by the composite. Mass transfer restriction effects could reduce the intensity of the amperometric signal. Comparing the amperometric response signals generated by the different composites proposed in this work, the range of analysis measure and the nature of the sample may be considered before the choice of the biosensor working electrode. A dilution is frequently necessary to fit the analyte sample concentration into the linear range sensibility of the developed biosensor. The reduction of waste discarded after each analysis contributed to the reduction of waste disposal in the environment since low sample volumes are used for the electrochemical analysis. This advantage encourages researchers in maintaining their efforts in developing continuously new composites and devices to improve the analytical proceedings [36].

3. Conclusions

Ethanol was estimated by electrochemical methods based on AOD- and HRP-modified electrodes with simple assembly and operation since it does not require the addition of a cofactor other than O_2. The mixture experimental design was efficient in determining the quantities of each component of the enzyme immobilization solution, avoiding spending time and the use of many expensive reagents to obtain the desired results. The graphite powder-silver nanoparticle-polyaniline was used as conducting composite for the working electrode, and the composite surface roughness was adequate to immobilize the enzymes AOD and HRP (solution composed of 85% bi-enzymatic solution, 10% albumin and 5% glutaraldehyde). The biosensor showed response signal linearity in the concentration range from 0 to 30 g/L (R^2 = 0.983) and sensitivity of 0.004 mA L/g, and LOD and LOQ were 3.54 and 11.8 g/L, respectively. These results indicate a robust bioelectrode suitable for analyses of contents up to 30 g/L of ethanol in samples. The repeatability and reproducibility of the biosensor were considered satisfactory since the variance and standard deviation showed low values. Relative standard deviation values were also low, below 2.2 and 0.9% for repeatability and reproducibility, respectively. The biosensor prepared with the AgNPs/GEC, working with the better bi-enzymatic solution composition, detected very low ethanol concentrations in a range of 0–0.35 g/L (R^2 = 0.984), a higher sensitivity of 6.899 μA.L/g, LOD of 3.48×10^{-3} g/L and LOQ of 0.0116 g/L. Two AgNPs composites are investigated in this work and characterized to develop bi-enzymatic biosensor for ethanol analysis. The ranges of ethanol solutions and the analytical performances were also investigated. Low levels of ethanol in standard samples could be detected with the AgNPs/GEC bi-enzymatic biosensor. The AgNPs/PANI/GEC biosensor was well fitted for high ethanol content in standard samples.

Acknowledgements

The authors would like to thank Leopoldo Américo Miguez de Mello, Petrobras Research and Development Center (CENPES) and National Council for Scientific and Technological Development (CNPQ-PIBIC)/Federal University of Rio de Janeiro (UFRJ), for the scholarship financial support and Toyobo of Brazil for the donation of *horseradish peroxidase* enzyme.

Author details

Eliana Mossé Alhadeff[1]*, Alfredo Jackson Telles Bosco[1], Caio Fragale Pastusiak[1], Thais Anjos Correia[1] and Ninoska Isabel Bojorge Ramirez[2]

1 Biochemical Engineering Department, Federal University of Rio de Janeiro, Rio de Janeiro, RJ, Brazil

2 Chemical and Petroleum Engineering Department, Federal Fluminense University, Niterói, RJ, Brazil

*Address all correspondence to: ema@eq.ufrj.br

References

[1] Léger C, Bertrand P. Direct electrochemistry of redox enzymes as a tool for mechanistic studies. Chemical Reviews. 2008;**108**(7):2379-2438. DOI: 10.1021/cr0680742

[2] Turner APF. Biosensors: Sense and sensibility. Chemical Society Reviews. 2013;**42**(8):3184-3196. DOI: 10.1039/c3cs35528d

[3] Ronkainen NJ, Halsall HB, Heineman WR. Electrochemical biosensors. Chemical Society Reviews. 2010;**39**(5):1747-1763. DOI: 10.1039/b714449k

[4] Rocha-Gaso M-I, Villarreal-Gómez L-J, Beyssen D, Sarry F, Reyna M-A, Ibarra-Cerdeña C-N. Biosensors to diagnose Chagas disease: A brief review. Sensors. 2017;**17**:2629. DOI: 10.3390/s17112629

[5] Munteanu F-D, Titoiu AM, Marty J-T, Vasilescu A. Detection of antibiotics and evaluation of antibacterial activity with screen-printed electrodes. Sensors. 2018;**18**(3):901-927. DOI: 10.3390/s18030901

[6] Çolak O, Arslan F. Amperometric biosensing of ethanol based on integration of alcohol dehydrogenase with a Pt/PPy–PVS/MB electrode. Turkish Journal of Chemistry. 2015;**39**:84-95

[7] Pires D, Bellissimo-Rodrigues F, Pittet D. Ethanol-based hand rubs: Safe for patients and health care workers. American Journal of Infection Control. 2016;**44**:858-859. DOI: 10.1016/j.ajic.2016.02.016

[8] Hooda V, Gahlaut VK, Hooda V. Alcohol quantification: Recent insights into amperometric enzyme biosensors. Artificial Cells, Nanomedicine, and Biotechnology: An International Journal. 2018;**46**(2):398-410. DOI: 10.1080/21691401.2017.1315426

[9] Kroutil W, Mang H, Edegger K, Faber K. Biocatalytic oxidation of primary and secondary alcohols. Advanced Synthesis & Catalysis. 2004;**346**(2-3):125-142. DOI: 10.1002/adsc.200303177

[10] Veitch N. Horseradish peroxidase: A modern view of a classic enzyme. Phytochemistry. 2004;**65**(3):249-259. DOI: 10.1016/j.phytochem.2003.10.022

[11] Azevedo A, Prazeres DMF, Cabral JMS, Fonseca LP. Ethanol biosensors based on alcohol oxidase. Biosensors and Bioelectronics. 2005;**21**(2):235-247. DOI: 10.1016/j.bios.2004.09.030

[12] Santana ACO, Southgate EF, Mendes JPBG, Dweck J, Alhadeff EM, Ramirez NIB. Characterization of an HRP-AOX-polyaniline-graphite composite biosensor. Journal of Electrochemical Science and Engineering. 2014;**4**(4):165-175. DOI: 10.5599/jese.2014.0057

[13] Stejskal J, Sapurina I. Polyaniline—A conducting polymer. In: Schubert U, Hüsing N, Laine RM, editors. Materials Syntheses: A Practical Guide. Vienna: Springer; 2008. pp. 199-207. DOI: 10.1007/978-3-211-75125-1_26

[14] Alhadeff E, Bojorge N. Graphite-composites alternatives for electrochemical biosensor. In: Cuppoletti J, editor. Metal, Ceramic and Polymeric Composites for Various Uses. Rijeka, Croatia; 2011. pp. 597-620. Available from: https://cdn.intechopen.com/pdfs-wm/16725.pdf

[15] Wang G, Wang W, Wu J, Liu H, Jiao S, Fang B. Self-assembly of a silver nanoparticles modified electrode and its electrocatalysis on neutral red. Microchimica Acta. 2009;**164**(1):149-155. DOI: 10.1007/s00604-008-0050-1

[16] Bosco AJT, Alhadeff EM, Mihos F, Das CSS, Yokoyama L, Santos VM, et al. Optimization of method of preparing carbon paste electrode. Chemical Engineering Transactions. 2015;**43**:2443-2448. DOI: 10.3303/CET1543408

[17] Alhadeff EM, Salgado AM, Pereira Jr N, Valdman B. A sequential enzymatic microreactor system for ethanol detection of gasohol mixtures. In: Davison BH, Evans BR, Finkelstein M, McMillan JD, editors. Twenty-Sixth Symposium on Biotechnology for Fuels and Chemicals. ABAB Symposium. Humana Press. Applied Biochemistry and Biotechnology. 2005;**121-124**:361-372. DOI: 10.1007/978-1-59259-991-2_31

[18] Massachusetts Department of Environmental Protection. Large Volume Ethanol Spills—Environmental Impacts & Response Options. 2011. 107 p. Available from: https://books.google.com.br/books?id=x5fPjwEACAAJ

[19] Ahammad AJS. Hydrogen peroxide biosensors based on horseradish peroxidase and hemoglobin. Journal of Biosensors and Bioelectronics. 2013;**S**:9. DOI: 10.4172/2155-6210.S9-001

[20] Ahmed MG, Charyulu RN, Harish NM, Prabhu P. Formulation and in-vitro evaluation of chitosan films containing tetracycline for the treatment of periodontitis. Asian Journal of Pharmaceutics. 2009;**3**(2):113-119. DOI: 10.4103/0973-8398.55048

[21] Turkevich J, Stevenson PC, Hillier J. A study of the nucleation and growth processes in the synthesis of colloidal gold. Discussions of the Faraday Society. 1951;**11**:55-75. DOI: 10.1039/DF9511100055

[22] Rashid MU, Bhuiyan MKH, Quayum ME. Synthesis of silver nano particles (Ag-NPs) and their uses for quantitative analysis of vitamin C tablets. Dhaka University Journal of Pharmaceutical Sciences. 2013;**12**(1): 29-33. Available from: http://www.banglajol.info/index.php/JPharma/article/view/16297/11546

[23] Kholoud MM, Abou E-N, Eftaiha A, Al-Warthan A, Reda AAA. Synthesis and applications of silver nanoparticles. Arabian Journal of Chemistry. 2010;**3**(3):135-140. DOI: 10.1016/j.arabjc.2010.04.008

[24] Power A, Betts AJ, Cassidy JF. The bench synthesis of silver nanostructures of variable size and an introductory analysis of their optical properties. Australian Journal of Education in Chemistry. 2013;**73**:1-15. Available from: http://www.thehealthwell.info/node/666532

[25] Krutyakov YA, Kudrinskiy AA, Yu Olenin A, Lisichkin GV. Synthesis and properties of silver nanoparticles: Advances and prospects. Russian Chemical Reviews. 2008;**77**(3):233-257. Available from: http://iopscience.iop.org/0036-021X/77/3/R03

[26] Mani U, Dhanasingh S, Arunachalam R, Paul E. A simple and green method for the synthesis of silver nanoparticles using *Ricinus communis* leaf extract. Progress in Nanotechnology and Nanomaterials. 2013;**2**(1):21-25. Available from: http://www.academicpub.org/PNN10013-20130416-090539-8405-11253.pdf

[27] Mocanu A, Pasca RD, Tomoaia G, Garbo C, Frangopol PT, Horovitz O, et al. New procedure to synthesize silver nanoparticles and their interaction with local anesthetics. International Journal of Nanomedicine. 2013;**8**(1):3867-3874. DOI: 10.2147/IJN.S51063

[28] Calderon-Jimenez B, Johnson ME, Bustos ARM, Murphy KE, Winchester MR, Baudrit JRV. Silver nanoparticles: Technological advances, societal impacts, and

metrological challenges. Frontiers in Chemistry. 2017;**5**:1-26. DOI: 10.3389/fchem.2017.00006

[29] de Prada AG-V, Peña N, Mena ML, Reviejo AJ, Pingarrón JM. Graphite-Teflon composite bienzyme amperometric biosensors for monitoring of alcohols. Biosensors and Bioelectronics. 2003;**18**(10):1279-1288. DOI: 10.1016/S0956-5663(03)00074-5

[30] Nunes GS, Jeanty G, Marty J-L. Enzyme immobilization procedures on screen-printed electrodes used for the detection of anticholinesterase pesticides: Comparative study. Anaytical Chimica Acta. 2004;**523**:107-115. DOI: 10.1016/j.aca.2004.03.100

[31] Silva VPA, Alves CR, Dutra RF, de Oliveira JE, Rondina D, Furtado RF. Biossensor amperométrico para determinação de peróxido de hidrogênio em leite. Eclética Química. 2011;**36**(2):143-157. Available from: http://www.scielo.br/pdf/eq/v36n2/a08v36n2.pdf

[32] Johansson K, Jönsson-Pettersson G, Gorton L, Marko-Varga G, Csöregi E. A reagentless amperometric biosensor for alcohol detection in column liquid chromatography based on co-immobilized peroxidase and alcohol oxidase in carbon paste. Journal of Biotechnology. 1993;**31**(3):301-316. DOI: 10.1016/0168-1656(93)90076-Y

[33] Skoog DA, Crouch SR, Holler FJ, editors. Principles of Instrumental Analysis. 6th ed. Thomson Brooks; 2014

[34] Ang LF, Por LY, Yam MF. Study on different molecular weights of chitosan as an immobilization matrix for a glucose biosensor. PLoS ONE. 2013;**8**(8):1-13. Available from: e70597. DOI: 10.1371/journal.pone.0070597

[35] Bard AJ, Faulkner LR. Electrochemical Methods: Fundamental and Applications. 2nd ed. Hoboken, NJ: John Wiley & Sons; 2001

[36] Ahmadalinezhad A, Chen A. High-performance electrochemical biosensor for the detection of total cholesterol. Biosensors and Bioelectronics. 2011;**26**(11):4508-4513. DOI: 10.1016/j.bios.2011.05.011

Chapter 10

Biosensors for Determination of Heavy Metals in Waters

Amra Odobašić, Indira Šestan and Sabina Begić

Abstract

Biosensors are nowadays a powerful alternative to conventional analytical techniques for controlling the quality of not only natural water but also process water used by the food industry during the production process, as well as wastewater prior to release into natural watercourses. The goal is to provide the required quality and safety of water from the standpoint of heavy metal contamination. The basic and most important characteristics of biosensors are high sensitivity, short response time, specificity, and relatively low production cost. Biosensors can detect the presence and measure the content of various toxic substances (pesticides, heavy metals, etc.) not only in water but also in food. Detection of contaminants, primarily heavy metals in water used in food production processes, is a potential area of biosensor application in the food industry. Biosensors can be adapted for direct and continuous (online) monitoring by measuring certain analytes that can affect the quality and safety of water. This chapter will give an overview of the development and application of biosensors in order to control the quality and safety of water from the standpoint of the presence of heavy metals.

Keywords: biosensors, heavy metals, natural water, waste water

1. Introduction

Monitoring of water pollution is very important for the preservation of the environment and prevention of negative impacts that it can have on human health. Therefore, great attention is paid to simplifying procedures for detection and monitoring of pollutants. Heavy metals are particularly dangerous due to their ability to accumulate over time in both plants and animals, as well as in water. For these reasons, there are already developed different methods that determine their concentrations generally in the environment.

Biosensors represent a simple, reliable, and fast solution for monitoring water pollution caused by various heavy metals. The small size of biosensor devices has enabled their in situ application, thus avoiding long-term and sometimes expensive measurements in laboratories.

According to IUPAC, biosensor represents a "self-contained integrated device, which is capable of providing specific quantitative or semi-quantitative analytical information using a biological recognition element (biochemical receptor) which is retained in a direct spatial contact with an electrochemical transduction element" [1, 2]. Biosensors allow not only determining the presence and overall biologically available concentrations of heavy metals in water but also assessing their biological effects, such as toxicity or cytotoxicity, which are sometimes more important than chemical composition information.

2. General characteristics of heavy metals

The term "heavy metals" refers to all metals except Al, Na, Ca, Mg, and K, i.e., to all metals that have a density higher than 5 g/cm^3. It includes a number of physiologically important elements such as Fe, Cu, Zn and Mn, then highly toxic Pb, As, Hg, Cd, Sb, Cr(VI) and less toxic Au, Ag, Mo, Cr(III) and Co [3]. The physiological and toxicological effects of these elements represent a collection of very different mechanisms.

Even at very low concentrations, they pose a threat to the environment and human health, because they are not biodegradable, so heavy metals are the cause of one of the most serious pollution problems. The most important nonessential heavy metals which affect the surface water systems are cadmium, chromium, mercury, lead, arsenic, and antimony [4].

Heavy metals present in pesticides and therapeutic agents are additional pollution sources. Burning of fossil fuels containing heavy metals and increasing industrial applications of metals such as metal galvanizing, paint and varnish industry, and mining and chemical industries are the main source of pollution of water systems by heavy metals.

Heavy metals are transported with waste water at the place of discharge and contaminate water sources downstream from an industrial site. In water, heavy metals have the ability to bind to the surface of microorganisms, from where they are transported inside the cell where they can be involved in chemical reactions and change chemically.

The majority of known techniques can determine the total amount of heavy metal ions. In addition, laboratory techniques that are routinely used for the analysis of metal ions, such as atomic absorption spectrometry, inductively coupled plasma mass spectrometry, anodic stripping voltammetry, and X-ray fluorescence spectrometry, require sophisticated equipment, pretreatment of samples, or qualified operators.

However, today it is known that only certain oxidation states of biologically available metal ions pose the greatest risk to human health and the environment. For example, "Cr(III) is an essential nutrient required in insulin action and sugar and fat metabolism, while Cr(VI) is believed to be highly toxic and carcinogenic" [5].

2.1 Mechanism of heavy metal toxicity

Metals and metalloid ions can be divided into three groups according to their toxicity. The first group includes metals (metalloids) that are toxic at extremely low concentration, such as lead, cadmium, and mercury. "Metals of the second group (arsenic, bismuth, indium, antimony and thallium) are less toxic, i.e., they are toxic only in higher concentrations. The third group includes metals (metalloids) of essential importance, such as copper, zinc, cobalt, selenium and iron, which are necessary for different chemical and biochemical processes in the body, and are toxic only above a certain concentration." Concentration window "of these heavy metals is somewhere between toxic and maximum permissible limits" [6].

Table 1 gives critical concentrations of some heavy metals in natural waters according to EPA [7].

Metal	Max. allowable concentration (μg/ml)
Mercury	0.002
Arsenic	0.5
Lead	0.5
Copper	0.6
Cadmium	0.04
Zinc	5

Table 1.
Critical concentrations of some heavy metals in natural waters according to EPA

The toxic effects of heavy metals can be the result of changes in numerous physiological processes at the cellular or molecular level caused by the inactivation of the enzyme. It can also occur as a result of the blocking of functional groups of metabolically important molecules or by replacing the essential elements and disturbing the integrity of the membrane. A rather frequent consequence of heavy metal poisoning is the production of reactive oxygen species (ROS) due to interference with the transport activities of electrons, especially the chloroplast membrane [8]. This increase in ROS exposes cells to oxidative stress that leads to peroxidation of lipids, biological damage of macromolecules, membrane decay, and DNA splitting [9].

They can penetrate into the organism in elemental form, in salt form, or as organometallic compounds, wherein the process of absorption, distribution, deposition, and elimination depends on the form in which the metal is present. Metals are very toxic because they are either in ionic form or within the compound, soluble in water, and easily absorbed by living organisms [3].

The mobility of heavy metals in water is particularly affected by the pH of water, the presence of hydrated forms of Mn and Fe, the concentration of carbonates and phosphates, as well as the content of organic matter. In addition, if the medium is very acidic and increased redox potential, the mobilization of Cu and Pb occurs, and under the reduction conditions, the hydroxides Mn and Fe are mobilized.

Heavy metals which are mostly the subject of research and monitoring in water and also generally in the environment due to their pronounced toxicity are arsenic, chromium, lead, mercury, and cadmium, while zinc, cobalt, copper, iron, and manganese are also interesting because they belong to the group of essential elements. The level of toxicity for some of these heavy metals is at or slightly above the concentration in which they are naturally found in nature [10]. Heavy metals occur in the environment naturally or as a result of human activities. Natural sources include volcanic eruptions, weathering (acid rock drainage), and discharge into rivers, lakes, and oceans.

Anthropogenic sources of heavy metals have emerged with the development of society. For example, the release of metal from the dishes causes contamination of food and water with metals.

2.1.1 Essential metals

2.1.1.1 Iron

Iron belongs to a group of essential metals and is crucial for a number of synthetic and enzyme processes in the human body. Most of the iron in our body exists as part of the hemoglobin molecule or myoglobin molecule. In addition to the vital importance it has for most living organisms, iron is potentially toxic at high concentrations. The effect of iron on aquatic organisms and their habitats is mostly indirect. Combined direct and indirect effects of contamination of the aquatic environment cause a decrease in biodiversity and number of fish. In aqueous solutions, the Fe^{3+} ion is in the form of the aqua complex, $Fe(H_2O)_6{}^{3+}$, which is quite hydrolyzed (hydrolysis starts at pH 1). Hydrolysis of Fe(III) ions depends on the type of ionic environment, temperature, and the presence of other substances. The results of the researches show that the most important chemical types are found in hydrolyzed solution.

2.1.1.2 Copper

Copper is a microelement of outstanding biological importance and is part of essential metabolic pathways. Copper ions play a key role in active centers of oxidoreductases, such as superoxide dismutase (Cu, Zn-SOD), [5], an enzyme important for maintaining a low level of free radicals in the cell, thus protecting biomolecules such as proteins and lipids from the pathological conditions.

Copper deficiency can cause anemia, because insufficient amount of copper causes poor absorption of iron, reducing the number of red blood cells. The lack of copper also reduces the amount of white blood cells and therefore the resistance of the organism to diseases. In general, copper is not considered to be a major ecotoxicological problem, but its widespread distribution and exposure to exhaust gases are certainly the reasons why copper is involved in the structuring of ecosystems. Copper is found in three oxidation states, Cu^+, Cu^{2+}, and Cu^{3+}, with the Cu^{2+} form being the most common. The most mobile forms of copper are Cu^{2+} and $CuOH^+$. In the aqueous environment, copper is found in three basic forms, as suspended, colloidal, and dissolved. The accumulation of copper in the aquatic environment results in the primary exposure of aquatic organisms. Aquatic organisms can accumulate dissolved copper by direct absorption through the body surface, while colloidal forms of this metal are introduced into the body by ingesting contaminated food.

2.1.1.3 Zinc

Zinc participates in the structure of many enzymes and is an essential element. It is attached to insulin and plays a significant role in the metabolism of nucleic acids and amino acids, DNA replication, and gene expression. However, like all other essential metals, zinc in higher concentrations is toxic to living organisms. Zinc can bioaccumulate in fish, and the degree of bioaccumulation usually depends on the exposure mode, as well as the conditions prevailing in the observed aquatic environment. Conditions that may affect the toxicity of zinc (but also other heavy metals) in the aquatic environment are the content of Ca and Mg, the pH of water, the content of the hydroxide (alkalinity), and the content of dissolved natural organic matter, i.e., humic substances.

2.1.1.4 Cobalt

The required amount of cobalt in the body is about 5 mg for vitamin B12 to avoid anemia. In general, cobalt has low toxicity. Gastrointestinal (digestive tract) absorption of soluble cobalt compounds is estimated to be 25%. However, cobalt is toxic to humans. When cobalt has been used as an additive in beer (for foam stabilization), severe biventricular heart failure and a high mortality rate were observed in heavy beer drinkers [11].

Long-term inhalation of cobalt dust irritates the respiratory tract and can cause chronic bronchitis, and cobalt salts can cause benign dermatosis. Cobalt occurs in oxidation states 0, +1, +2, +3, and +4, and most of its compounds have an oxidation number +2 and +3, of which the cobalt(II) compounds are more stable. Most cobalt(II) compounds have an ionic character (halides and numerous Co(III) complexes). Cobalt is relatively a nonreactive metal. It does not oxidize under dry and humid conditions at normal temperatures. It binds to halogen elements by heating. Cobalt is used in the production of artificial fertilizers and so can be found in higher concentrations in soil and water. It is also used in medicine, in the treatment of anemia that cannot be treated with iron.

2.1.2 Toxic metals

2.1.2.1 Lead

Lead in the environment mainly comes from anthropogenic sources such as combustion of fossil fuels, landfills and fires at landfills, waste industrial sludges, phosphate-based fertilizers, pesticides, and exhaust gases from vehicles.

It is found in the form of sulphates, sulphides, and carbonates. It is considered the leading environmental pollutant and is increasingly endangering the living world, especially the surrounding areas of large industrial plants, frequent roads, and large cities.

The intensity of the adoption of lead depends on its concentrations in soil, soil pH, organic matter content, ratio of cations and anions, and other environmental factors. Human is exposed to toxic effects of lead by consuming food and water that are contaminated with this heavy metal but also by inhaling particulate matter with lead content. Absorption over the skin is only possible for tetraethyl and tetramethyl lead. Lead is rapidly absorbed into the bloodstream and binds to red blood cells in the form of Pb^{2+}, and via blood about 90% is deposited in the bones in the form of $Pb_3(PO_4)_2$. In the case of acidosis (increased acidity), the mobility of lead from the bones in the form of Pb^{2+} which has a toxic effect on the central nervous, circulatory, and immunological systems and kidneys can occur. [10]

2.1.2.2 Mercury

Mercury vapors and organic compounds of mercury are very strong poisons. Harmful substances are released by combustion of fossil fuels, and the risk of pollution threatens also due to increased use of mercury in industry and agriculture [12].

2.1.2.3 Chromium

In its compounds, chromium exists in several oxidation states: from bivalent to hexavalent. In solutions, chromium can occur in trivalent and hexavalent forms. Hexavalent chromium is usually present in the compounds as chromate $(CrO_4)^{2-}$ or dichromate $(Cr_2O_7)^{2-}$ ion. Cr(VI) is toxic due to its high degree of oxidation and easily enters the biological membranes. Therefore, this form of chromium is considered carcinogenic. Because chromium(VI) is toxic, carcinogenic, and mutagenic to living organisms, damages the liver, and causes lung congestion, skin irritation, and the formation of ulcer, it needs to be removed from the wastewater before their release into natural recipients. On the other hand, trivalent chromium, Cr(III), is 300 times less toxic than chromium(VI). Chromium is a vital nutrient for many animal and plant species, but it can also cause allergic reactions on the skin and can be carcinogenic [13].

3. Biosensors

A biosensor is an analytical device consisting of immobilized biological material in direct contact with a compatible transducer that will convert the biochemical signal into a measurable electrical signal. Biomolecules are responsible for specific recognition of the analyte, while the physicochemical converter provides electrical output signal that is amplified by electronic component [14]. Biosensors find application in various areas, from agriculture, food quality control, medicine, army, and control of various processes in the environment. Biosensors can provide quick information about the site of pollution, which is necessary for environmental control and monitoring. In addition, the advantage of biosensors over other analytical methods is their mobility that allows researchers to measure the in situ pollutant concentration and the ability to measure the concentration of pollutants in situ without additional sample preparation. Also, in addition to the determination of specific compounds, they can provide information on their biological effect (e.g., toxicity of a compound).

Due to exceptional performances, including high specificity and sensitivity, rapid response, low cost, relatively small size, and simple operation, biosensors have become an important tool for detecting chemical and biological components and their monitoring for clinical, nutritional, and ecological needs [15].

3.1 General characteristics of biosensors

Biosensors are analytical sensory devices that combine physical and chemical sensing techniques [16, 17]. Their performance is based on direct contact of two elements: biological and physicochemical, whose tight bond is achieved by physical or chemical methods of immobilization. Biological element serves as a receptor (bioreceptor), i.e., for the recognition of particular analyte from the medium of interest, based on the interaction of analyte and bioreceptor. Physicochemical transducer converts the response that occurs as a result of analyte-bioreceptor interaction on their interface into a measurable signal which can be processed and displayed in the form of readable values. For proper biosensor operation, the biological compound has to be immobilized in the vicinity of the transducer, and immobilization can be done either by physical entrapment or chemical attachment. Only small amounts of bioreceptor molecules are required, and they will be repeatedly used for measurements [18].

The displayed values are in correlation with the detected analyte-bioreceptor interactions, i.e., the concentration of a specific analyte or group of analytes in the analyzed sample [4, 16, 17]. General working principle of biosensors is illustrated in **Figure 1**.

Although widely used, conventional analytical techniques require sophisticated instruments and highly trained personnel to conduct operational procedures and

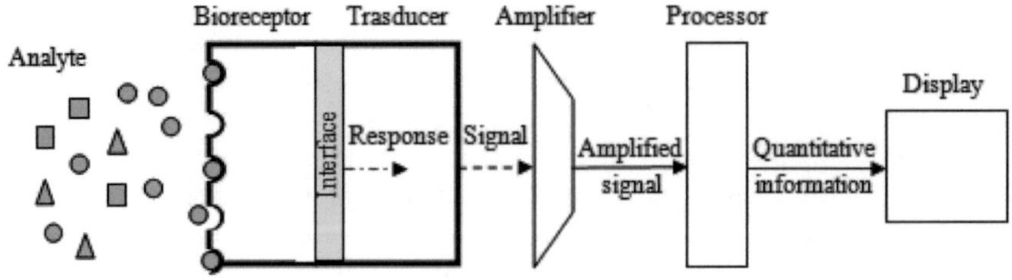

Figure 1.
Schematic illustration of a biosensor general working principle.

sample preparation, which makes them expensive and time-consuming [19, 20], thus not enabling determination of a large number of samples in a short time [21].

The main advantages of biosensors in relation to conventional analytical techniques are possibility of miniaturization and portability of device, reduced requirements for laboratory skills, reduced sample volume and pretreatment [1, 22], assessment of all possible types of analytes, inorganic or organic [23, 24], and possibility of performing single measurements or continuous real-time monitoring of analytes [1, 25]. Biosensors allow estimation of biological effects, e.g., toxicity of specific chemicals, because they can be used to detect their bioavailable concentrations [26].

3.2 Classification and types of biosensors

Biosensors can be divided into classes according to different approaches, among which the two are commonly used—type of biorecognition element (biocomponent, bioreceptor) and type of transduction system in biosensor. Each class of biosensors can be further classified into subclasses (**Figure 2**).

3.2.1 Classification by type of transducer

Based on the principle used in transduction systems, electrochemical, optical, piezoelectric, and thermal biosensors may be distinguished.

3.2.1.1 Electrochemical biosensors

The first proposed and commercialized biosensors were electrochemical biosensors, which is why they are most commonly reported. The basic principle of this class of biosensors is that the interaction between the biomolecule (bioreceptor) and the target analyte results in a chemical reaction that produces or consumes ions or electrons and in turn changes the electrical properties of the analyte solution, such as electrical current or potential. Transducer detects these changes by producing an electrochemical signal which is correlated with the amount of analyte present in the sample solution.

Advantages of electrochemical biosensors include minimal requirements for sample preparation and sensitivity at small sample volumes. It is also possible to perform sample analysis directly, which enables automation. Drawbacks of detections are poor reproducibility and stability [27].

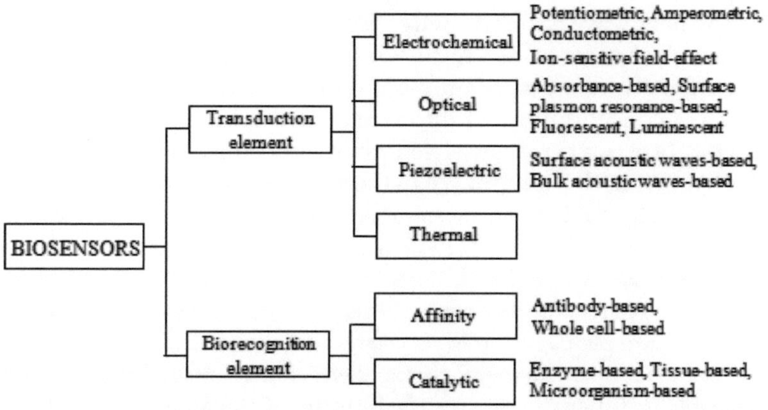

Figure 2.
Schematic illustration of the common classification of biosensors.

Electrochemical biosensors are classified according to the type of measured signal into subclasses: potentiometric, amperometric, conductometric, and biosensors based on ion-selective field-effect transistors (ISFETs). Different measurement principles always require a specific design of an electrochemical cell [21].

Potentiometric biosensors are based on the use of ion-selective electrode (ISEs) at the top of which an ion-selective membrane is placed which is responsible for selectivity to target ions in the presence of interfering ions in the sample. These devices measure the difference between the potential of the working and reference electrodes at essentially zero current, and this difference corresponds to the concentration of the analyte.

Amperometric biosensors are the most widespread class of electrochemical biosensors. Amperometric biosensors are more sensitive and faster than potentiometric but have poor selectivity and are susceptible to the interference of electroactive species that are not of interest [22, 28].

Conductometric biosensors are based on measurement of electrical conductivity in sample solution between two electrodes, as a consequence of the biochemical reaction. Conductometric biosensors operate at sufficiently low driving voltage, are not sensitive to light, do not require the use of a reference electrode, and can be produced using inexpensive technology [23, 24].

Biosensors based on the ion-selective field-effect transistors (ISFET) are the fourth class of electrochemical biosensors, suitable for the direct detection of ions. Change of activity of ions of a sample causes a change in the potential of the gate electrode that is brought into contact with the analyte solution. The change of the electric potential is then measured.

3.2.1.2 Optical biosensors

Optical biosensors are a biosensor class in which the transducer detects optical changes in the input light resulting from the interaction of the bioreceptor and the target analyte, and the amplitude of these changes is in correlation with the concentration of the present analyte in the analyzed sample. Among the significant advantages of these optical devices are insensitivity to electromagnetic interference, small instrumentation, simplicity, and noninvasiveness of measurement, as well as the possibility of application in vivo, since they are non-electrical biosensors. According to the optical configuration, biosensors can be intrinsic or extrinsic. In intrinsic biosensors, the incident light wave is closed in a wave guide or an optical fiber, along which it propagates, but the design of the structure in which the wave is closed is such that it allows the interaction of the wave with the analyte. In extrinsic biosensors, the light wave passes directly through the sample phase and reacts with it, and the optical fiber serves as a means of transmitting the signal.

Absorption-based biosensors are simple and inexpensive devices that allow the determination of concentrations of different analytes, based on the fact that each type of analyte absorbs a certain wavelength of light emitted into the sample. Guiding the light from the light source to the sample and from the sample to the detector can be performed using the same optical fiber or different fibers [29].

Surface plasmon resonance (SPR) biosensors use an optical detection technique where on the interface of metal and dielectrics, the amplified incident light hits the metal surface and excites the electrons, thereby generating electromagnetic waves (plasmons). Plasmon propagation is very sensitive to the changes in the refractive index of the material near the metal surface, which are caused by biomolecular interaction, such as, for example, specific binding of the analyte [30, 31].

Fluorescence-based optical biosensors can directly detect target atoms or molecules by measuring the change in the frequency of electromagnetic radiation emitted by them. The frequency change is stimulated by the absorption of radiation and the consequent appearance of the excited state of the target species. Detection can also be carried out indirectly, using fluorescence labels or fluorescence energy transfer (FRET).

Luminescence-based biosensors can be classified into chemiluminescent and bioluminescent optical biosensors. Unlike biosensors based on fluorescence, in these sensor devices, the triggered state of the target atoms or molecules is obtained as a result of their exothermic chemical reaction, and while returning to the ground state, the excited species emit light without or with minimal heat. When such a chemical reaction occurs within a biological organism, then it is a bioluminescence.

3.2.1.3 Piezoelectric biosensors

Piezoelectric biosensors are devices in which the biorecognition element is integrated with a piezoelectric material used as a transducer. Among many types of natural and synthetic materials that exhibit a piezoelectric effect, quartz crystals are most commonly used [28, 32] because of their availability, as well as high temperature resistance and chemical stability in aqueous solution. The basic principle of measurement for this type of biosensor is based on the ability of a piezoelectric material to generate electrical potential when deformed under the applied mechanical stress, and vice versa, to elastically deform when exposed to an electric field.

3.2.1.4 Thermal biosensors

Thermal biosensors, also called calorimetric or thermometric, are a biosensor class in which the transducer detects interactions between bioreceptors and analyte resulting in a change of temperature, which is in correlation with the concentration of the analyte. As thermal transducers in these devices, thermistors or thermopiles are used [21, 33]. Some of the advantages of thermal biosensors are detection without the need for labeling of reactants, not requiring frequent recalibration, and no disturbances by electrochemical and optical properties of the sample [21, 34]. In most research papers published about this type of sensor, described experiments were carried out using enzyme-based thermal biosensors, due to the exothermic nature of the reactions catalyzed by enzymes.

3.2.2 Classification by type of biocomponent/bioreceptor

Biocomponent/bioreceptor is responsible for the detection and interaction with the analyte and therefore is a very important part of any type of biosensor. The receptor is responsible for the selective and sensitive recognitions of the analyte, and the energy liberated during the interaction of the analyte and the receptor is converted into an electrical signal that is suitable for measurement. The most commonly used biological elements are enzymes and antibodies. Biosensors can be divided into two main categories: biocatalytic and affinity sensors based on the interaction between biological material and analyte.

Biocatalytic biosensors, also known as metabolism sensors, comprise a biological component that catalyzes the chemical conversion of the analyte with which it interacts and detect the magnitude of the resulting changes such as product formation, reactant disappearance, or inhibition of the reaction, which are correlated with the concentration of the analyte [35]. Affinity biosensors are based on selective interaction between the analyte and the biological component through their irreversible binding, resulting in a physicochemical change detected by the converter.

3.2.2.1 Antibody-based biosensors

Antibodies are proteins produced by the immune system in response to a foreign substance in the body. Also known as immunoglobulins (Ig), they are Y-shaped proteins generated by a type of white blood cells called B lymphocytes (B cells). Their ability to recognize specific molecules makes them suitable for use as biorecognition component in biosensors. During the process of biological recognition, the antibodies bind tightly to antigens forming complexes. There are five classes of antibodies, based on their structure and function: IgA, IgD, IgE, IgG, and IgM. Among them, IgG is the class most frequently used for heavy metal detection, because of their higher affinity and specificity compared to other classes. Antibodies such as monoclonal, polyclonal, or recombinant can be utilized in biosensors. Monoclonal antibodies are homogeneous antibodies, derived from single B cell; thus they all have the same specificity, i.e., to bind to one unique epitope (binding site) on a specific antigen. Unlike monoclonal antibodies, polyclonal antibodies are produced from different B cells against the same antigen and therefore have affinity for various binding sites of that antigen. This feature of polyclonal antibodies results in their stronger binding to the target species, but due to the recognition of multiple epitopes, they have higher potential for cross-reactivity, i.e., specificity for nontargeted antigens with similar structural regions as the targeted one. The production of recombinant antibodies is enabled by genetic engineering. Important properties of antibodies for providing accurate results for detection and measurement using biosensors are high sensitivity and specificity, with minimal cross-reactivity [36].

Different types of approaches have been developed and used for immobilization of Abs onto a sensor surface, such as covalent binding, non-covalent immobilization, and coupling by affinity interactions, because the immobilization is the crucial step which can affect the optimal performance of an antibody-based biosensor [37]. Reaction conditions, such as temperature, pH, and ionic strength, can also affect the activity of the antibodies [38].

3.2.2.2 Enzyme-based biosensors

Enzymes are biocatalysts that catalyze chemical reactions. Their task is to translate the characteristic substance (substrate) into a product. Enzymes are highly selective for the particular substrate which makes them suitable sensor material. Detection mechanism of enzyme-based biosensors is based on activation or inhibition of their activities as a response caused by heavy metals. Usually the metal ion reacts with the thiol groups present in enzymatic structures that result in conformational changes and thus affect the catalytic activity. Different enzymes have been used for the structure of biosensors based on inhibition. Enzymes such as glucose oxidase, urease, glutathione S-transferase, alkaline phosphatase, lactate dehydrogenase, acid phosphatase, and invertase have been utilized to detect metals such as cadmium, lead, copper, mercury, zinc, etc. However, inhibition-based biosensors have an important disadvantage, which is insufficient selectivity because some of the enzymes simultaneously inhibit several metals.

Biosensors based on immobilized enzymes are also used, and they show several advantages compared to free enzymes:

- A thousand times lower consumption of immobilized enzymes.
- Reduced interferences in differential mode.
- No preincubation is required.

- Faster analysis, less than 5 min.
- In the case of reversible inhibition, sometimes reactivation of the enzyme activity is not necessary.

The problem with biosensors based on enzymatic inhibition is that only a few enzymes are sensitive to heavy metals.

3.2.2.3 Protein-based biosensors

Proteins, such as phytochelatins or metallothioneins, can be used as biological components in biosensors when immobilized on the surface of the transducer [39]. The interaction of proteins and metals in the biosensor is realized through the formation of complexes, and the detection technique does not require labeling. The resulting changes in the protein layer are detected by measuring the electrical capacity or impedance by the relevant transducer. Using the protein biosensor enabled the assessment of bioavailable concentrations of heavy metals. In addition, using capacitive sensors, which belong to the class of electrochemical biosensors, it is possible to achieve much higher sensitivity to low concentrations of heavy metals, compared to cell-based devices.

3.2.2.4 Whole cell-based biosensors

Whole cell-based biosensors are based on using biosensing cells, such as microorganisms, plant cells, algae, fungi, protozoa, etc., which can be natural or recombinant [40]. The use of whole cells as biological elements of recognition has many advantages. Whole cell-based biosensors are usually cheaper than biosensors based on enzymes, because the whole cells can be easily cultivated and are easier to isolate and purify compared with enzymes. Whole cells are more tolerant to a significant change in pH, temperature, or ionic strength. A multistep reaction is possible because one cell can contain all the enzymes and cofactors needed to detect the analyte. Biosensors of this type can easily be regenerated or maintained by allowing cells to regrow while working in situ. Preparation of samples is usually not necessary. Compared to enzyme-based biosensors, the disadvantages of these devices are that they are susceptible to interference of contaminants that are not targeted analytes. They also have a relatively slow response, compared to other types of biosensors.

3.3 Application of biosensors in detection and monitoring of heavy metals

The unique biosensor features make them widely applicable in the field of water quality control, from the point of view of detecting and determining the concentration of heavy metals. The use of biosensors for individual or continuous measurements is dependent on the type of biologically active element. Since biological compounds such as cholesterol, glucose, urea, etc. are generally not electroactive, the combination of reactions is needed for obtaining an electroactive element, which leads to a change of current intensity [41]. **Table 2** shows the classification of biosensors based on the recognition component that was utilized for the detection of heavy metals.

A proper immobilization of the biosensing element onto the transducer surface maintains biomaterial functionality while ensuring accessibility of the receptor cells toward analytes and proximity of the bioreceptor and transducer. The factors which determine the choice of a suitable physical or chemical immobilization method

Type of bioreceptor	Analyzed heavy metal	Reference
2A81G5	Cd	[42]
Antibody ISB4	Cd	[43]
12F6	U	[44]
Alkaline	Zn	[45]
Phosphatase	Hg	[46]
Pyruvate enzymes	Cd	[47]
Oxidase	Hg	[48]
Urease	Hg, Ag	
Glutathione S-transferase	Cd, Zn	[49, 50]
Mer R proteins	Hg, Cu, Cd, Zn, Pb	[51, 52]
Metallothionein	Cd, Zn, Ni	[53]
Whole cells and cardiac cells	Hg, Pb, Cd, Fe, Cu, Zn	[54]

Table 2.
Classification of biosensors based on the recognition component that was utilized for the detection of heavy metals

are physicochemical properties of the analyte, nature of the chosen biosensing element, the type of used transducer, and the operating conditions of biosensor. Antibody-based biosensors can be used as an alternative approach for the detection of metal ions, due to antibody features such as high specificity and binding affinity for antigens harmful for the organism. Detection mechanism of these devices is based on antibody-metal ion complex formation. The resulted response of their immunochemical interaction is converted by a transducer to measurable values and processed to readable values. Antibodies are capable for antigen detection in very low concentrations [38], but if their cross-reactivity is high, they can yield false-positive results of an assay of heavy metals in water [55].

A monoclonal antibody that recognizes 16 different metal-EDTA complexes has been produced and evaluated in terms of its binding affinity. The obtained results showed that the antibody has a maximum binding affinity for cadmium and mercury-EDTA complexes. [56]. In the inhibition immunoassay where the measurement of Cd^{2+} in water samples was carried out using monoclonal antibodies firmly bound to the cadmium-EDTA complex, but not to EDTA without metal [42], the biosensor showed satisfactory insensitivity to cations Ca^{2+}, Na^{2+}, and K^{1+} it encountered and achieved a reliable measurement in the presence of 1 mM of excess Fe^{3+}, Mg^{2+}, and Pb^{2+}.

Monoclonal antibodies were used to detect Pb^{2+} without labeling, in a localized surface plasmon resonance-based optical biosensor [57]. The results of the experiment showed that at optimal monoclonal antibody immobilizing conditions, absorbability increased to 12.2% for detecting 10–100 ppb Pb(II)-EDTA complex with a limit of detection of 0.27 ppb.

Kulkarni et al. were the first to develop acid phosphatase-based fluorescence biosensor for the analysis of heavy metal ions Hg^{2+}, Cr^{2+}, and Cu^{2+}. Increased concentration of metal ions resulted in increased enzyme inhibition and therefore decreased fluorescence. The enzyme was stable for more than 2 months at 4°C [58]. They also observed that mixture of heavy metal ions exhibit positive effect on the performance of biosensor.

The urease enzyme has been widely investigated as a possible biocomponent in heavy metal detection biosensors. Urease has been tested single and in combination with other enzymes. Electrochemical biosensor based on urease and glutamic dehydrogenase (GLDH) was developed for detecting heavy metals in water samples [59]. Also, a disposable potentiometric biosensor based on pure urease was developed, with the ability to detect copper and silver at sub-ppm level. For the detection of Pb and

Cd in liquid samples, biosensors based on the combination of urease and acetylcholinesterase (Ache) were developed as a biocomponent with a detection limit of 1 ppb in water samples. It is known that ions of heavy metals inhibit alkaline phosphatase which was used for forming the biosensor with alkaline phosphatase as a biocomponent. It was found that the sensitivity of the developed biosensor to Cd^{2+} and Zn^{2+} was 10 ppb, whereas, with regard to ion Pb^{2+}, there was no significant inhibition.

Two protein-based biosensors were developed on the basis of GST-SmtA and MerR [60] proteins, and their sensitivity and selectivity for heavy metal ions (Cd^{2+}, Cu^{2+}, Hg^{2+}, and Zn^{2+}) were measured using a capacitance transducer. Both types of biosensors have shown high sensitivity, enabling detection of metal ions up to femtomolar concentration.

Capacitance protein-based biosensor using synthetic phytochelatins (ECs) was developed for the detection of heavy metal ions (Cd^{2+}, Cu^{2+}, Hg^{2+}, Pb^{2+}, and Zn^{2+}), and the results of the experiments showed a lower sensitivity for all metal ions except for Zn^{2+} compared to systems based on SmtA and MerR, which can be explained by conformational changes in the protein, taking into account that the change in capacitance is function of the resulting change in protein conformation [51].

In cell-based biosensors, bioelement is fused with reporter gene. The detection mechanism is based on the activation of the reporter gene upon the contact between bioreceptor and target analyte, yielding an output measurable signal that is a correlation with bioavailable concentration of heavy metal.

Various cell-based biosensors have been used for the detection of heavy metals in water due to their ease of production and field testing, the ability to perform fast single measurement, as well as continuous measurements, and the ease of identifying bioavailable concentrations of toxicants that allows estimation of effects that heavy metals have on living organisms.

The advantage of bacterial cells is resistance to environmental conditions that could destroy the sensory element if exposed to them, supplying it with a relatively stable environment. Due to specific metabolic pathways used in microorganisms, compared to isolated enzymes, microbial sensors have the potential for more selective analysis of heavy metals which cannot be measured by simple enzyme reactions [61].

In order to be available for any sensing mechanism that is based inside the cell, there is a need for analytes to be able to enter the cell via diffusion, nonspecific uptake, or active transport. Alternative approaches are implemented in the cases when membrane permeability for an analyte is not sufficient. These approaches include allocation of the recognition element to the outside of the cell or the introduction of an appropriate transport mechanism for importing the analyte [61].

A large number of studies in which performances of whole cell-based biosensors were tested have utilized electrochemical and optical transducers. For detection of heavy metal ions (Cd^{2+}, Cu^{2+}, Fe^{3+}, Hg^{2+}, Pb^{2+},and Zn^{2+}) at concentrations of 10μM, a mammalian heart cells-based biosensor was developed [54], with excellent performance in terms of frequency selection, amplitude and duration of detection within 15 min.

Biosensor, based on immobilized engineered bacteria *Alcaligenes eutrophus* (AE1239) and optical transducer, was utilized for monitoring the bioavailable copper ions in synthetic water samples, wherein the lowest limit of detection was 1 μM [62].

4. Conclusions

Biosensors have a very wide range of applications, from environmental monitoring, food safety, detection of various diseases, use in artificial implantable devices such as pacemakers to the detection of drugs.

Application for pollution monitoring requires the biosensor to work from several hours to several days. Such biosensors are a tool for "long-term monitoring." Whether it is a long-term follow-up or analysis of individual shots, biosensors are used as technologically advanced devices both in settings with limited resources and in sophisticated medical settings.

Considering the complex and critical situation in the field of environmental protection, and the state of natural waters from the aspect of pollution with heavy metals, and taking into account the toxicity of heavy metal ions, it is necessary to continuously work on finding new efficient techniques for their detection. Conventional analytical techniques can no longer satisfy the needs of constant monitoring and frequent field analysis of water because they are expensive, often with bulky equipment and a long analysis time, and require well-trained analysts. Biosensors can be used to overcome the limitations of conventional methods. In the future, designing a biosensor with the appropriate material will surely help in the selective identification of metal ions not only from water but also from any other matrix.

Author details

Amra Odobašić[1]*, Indira Šestan[1] and Sabina Begić[2]

1 Faculty of Technology, Department of Physical and Electrochemistry, University of Tuzla, Tuzla, Bosnia and Herzegovina

2 Faculty of Technology, Department of Chemical Technology, University of Tuzla, Tuzla, Bosnia and Herzegovina

*Address all correspondence to: amra.odobasic@untz.ba

References

[1] Pearson JE, Gill A, Vadgama P. Analytical aspects of biosensors. Annals of Clinical Biochemistry. 2000;**37**(2):119-145

[2] Thevenot DR, Toth K, Durst RA, Wilson GS. Electrochemical biosensors: Recommended definitions and classification. Biosensors and Bioelectronics. 2001;**16**(1-2):121-131

[3] Šestan I. Investigation kinetics of the reaction of the formation of metallic complexes and the biogeodistribution of heavy metals in the system sediment/water [Doctoral thesis]. University of Tuzla; 2015

[4] Gumpu MB, Sethuraman S, Krishnan UM, Rayappan JBB. A review on detection of heavy metal ions in water—An electrochemical approach. Sensors and Actuators B. 2015;**213**:515-533

[5] Abdou KA, Khadiga IA. Biosensors for monitoring of the environmental heavy metals. Journal of Scientific and Engineering Research. 2016;**3**(3):31-38. ISSN: 2394-2630 CODEN(USA): JSERBR

[6] Odobašić A. Specification of heavy metals by polarographic techniques in natural waters [Doctoral thesis]. Faculty of Technology, University of Tuzla; 2005

[7] Mills WM, et al. Part I (Revise 1995), EPA 600/6-85/002a. Washington, DC 20460: USA EPA Office of Research and Development

[8] Pagliano C et al. Evidence for PSII-donor-side damage and photoinhibition induced by cadmium treatment on rice (*Oryza sativa* L.). Journal of Photochemistry and Photobiology B: Biology. 2006;**84**:70-78

[9] Quartacci MF, Cosi E, Navari-Izzo F. Lipids and NADPH-dependent superoxide production in plasma membrane vesicles from roots of wheat grown under copper deficiency and excess. Journal of Experimental Botany. 2001;**152**:67-75

[10] Đozić A, Hodžić I. The Presence of Heavy Metals in Surface and Industrial Wastewater in the Municipality of Lukavac. Tuzla: Centre for Ecology and Energy; 2016

[11] Packer M. Cobalt cardiomyopathy: A critical reappraisal in light of a recent resurgence. Circulation Heart Failure. 2016;**9**(12):1-10. DOI: 10.1161/CIRCHEARTFAILURE.116.003604

[12] Cargnelutti D, Almeri Tabaldi L, Spanevello RM, de Oliveira Jucoski G, Battisti V, Redin M, et al. Mercury toxicity induces oxidative stress in growing cucumber seedlings. Chemosphere. 2006;**65**:999-1006

[13] Cabatingan LK et al. Potential of biosorption for the recovery of chromate in industrial wastewaters. Industrial and Engineering Chemistry Research. 2001;**40**:2302-2309

[14] Scheller F, Schubert F. Biosensors. New York: Elsevier Science Publishers; 1992

[15] Amine A, Mohammadi H, Bourais I, Palleschi G. Enzyme inhibition-based biosensors for food safety and environmental monitoring. Biosensors & Bioelectronics. 2006;**21**:1405-1423

[16] Najeeb J, Ali J, Ali MA, Aslam MF, Raza A. Biosensors: Their fundamentals, designs, types and most recent impactful application: A review. Journal of Biosensors and Bioelectronics. 2017;**8**:235

[17] Bhalla N, Jolly P, Formisano N, Estrela P. Introduction to biosensors. Essays in Biochemistry. 2016;**60**:1-8

[18] Wilson JS. Sensor Technology Handbook. Vol. 1. Newnes: Elsevier; 2005. ISBN: 0-7506-7729-5

[19] Busairi N, Syahir A. Recent advances in mercury detection: Towards enabling a sensitive and rapid point-of-check measurement. Journal of Toxicology and Risk Assessment. 2018;**4**(1):1-10

[20] Mishra GK, Barfidokht A, Tehrani F, Mishra RK. Food safety analysis using electrochemical biosensors. Food. 2018;**7**:141

[21] Wang Y, Xu H, Zhang J, Li G. Electrochemical sensors for clinical analysis. Sensors. 2008;**8**:2043-2081

[22] Martinkova P, Kostelnik A, Valek T, Pohanka M. Main streams in the construction of biosensors and their applications. International Journal of Electrochemical Science. 2017;**12**:7386-7403

[23] Dzyadevych SV, Jaffrezic-Renault N. Conductometric biosensors. In: Schaudies RP, editor. Biological Identification. Vol. 6. Cambridge, UK: Woodhead Publishing; 2014. pp. 153-193

[24] Jaffrezic-Renault N, Dzyadevych SV. Conductometric microbiosensors for environmental monitoring. Sensors. 2008;**8**:2569-2588

[25] Sang S, Wang Y, Feng Q, Wei Y, Ji J, Zhang W. Progress of new label-free techniques for biosensors: A review. Critical Reviews in Biotechnology. 2016;**36**(3):465-481

[26] Damborsky P, Svitel J, Katrlik J. Optical biosensors. Essays in Biochemistry. 2016;**60**:91-100

[27] Gautam P, Suniti S, Prachi K, Amrita D, Madathil B, Nair AN. A review on recent advances in biosensors for detection of water contamination. International Journal of Environmental Sciences. 2012;**2**(3):1565-1574

[28] Monosik R, Stredansky M, Sturdik E. Biosensors—Classification, characterization and new trends. Acta Chimica Slovaca. 2012;**5**(1):109-120

[29] Bosch ME, Sanchez AJR, Rojas FS, Ojeda CB. Recent development in optical fibre biosensors. Sensors. 2007;**7**:797-859

[30] Wijaya E, Lenaerts C, Maricot S, Hastanin J, Habraken S, Vilcot J-P. Surface plasmon resonance-based biosensors: From the development of different SPR structures to novel surface functionalization strategies. Current Opinion in Solid State and Materials Science. 2011;**15**:208-224

[31] Asal M, Ozen O, Sahinler M, Polatoglu I. Recent developments in enzyme, DNA and immuno-based biosensors. Sensors. 2018;**18**:1924

[32] Pohanka M. The piezoelectric biosensors: Principles and applications, a review. International Journal of Electrochemical Science. 2017;**12**:496-506

[33] Perumal V, Hashim U. Advances in biosensors: Principle, architecture and applications. Journal of Applied Biomedicine. 2014;**12**:1-5

[34] Mohanty SP, Kougianos E. Biosensors: A tutorial review. IEEE Potentials. 2006;**25**(2):35-40

[35] Marazuela MD, Moreno-Bondi MC. Fiber-optic biosensors—An overview. Analytical and Bioanalytical Chemistry. 2002;**372**:664-682

[36] Ferrigno PK. Non-antibody protein-based biosensors. Essays in Biochemistry. 2016;**60**:19-25

[37] Sharma S, Byrne H, O'Kennedy RJ. Antibodies and antibody-derived analytical biosensors. Essays in Biochemistry. 2016;**60**:9-18

[38] Dzantiev BB, Zherdev AV. Antibody-based biosensors. In: Portable Biosensing of Food Toxicants and Environmental Pollutants. Boca Raton, FL: CRC Press Book; 2013. pp. 161-196

[39] Cornelis R, Crews H, Caruso J, Heumann K. Handbook of Elemental Speciation: Techniques and Methodology. Chichester, ENG: John Wiley & Sons, Ltd; 2003. ISBN: 0-471-49214-0

[40] Gu MB, Mitchell RJ, Kim BC. Whole-cell-based biosensors for environmental biomonitoring and application. Advances in Biochemical Engineering/Biotechnology. 2004;**87**:269-305

[41] Arora P, Sindhu A, Dilbaghi N, Chaudhury A. Biosensors as innovative tools for the detection of food borne pathogens. Biosensors and Bioelectronics. 2011;**28**:1-12

[42] Khosraviani M, Pavlov AR, Flowers GC, Blake DA. Detection of heavy metals by immunoassay: Optimization and validation of a rapid, portable assay for ionic cadmium. Environmental Science and Technology. 1998;**32**:137-142

[43] Blake DA, Jones RM, Blake RC, Pavlov AR, Darwish IA, Yu H. Antibody-based sensors for heavy metal ions. Biosensors and Bioelectronics. 2001;**16**:799-809

[44] Melton SJ, Yu H, Williams KH, Morris SA, Long PE, Blake DA. Field-based detection and monitoring of uranium in contaminated groundwater using two immunosensors. Environmental Science & Technology. 2009;**43**:6703-6709

[45] Satoh I. An apoenzyme thermistor microanalysis for zinc(II) ions with use of an immobilized alkaline phosphatase reactor in a flow system. Biosensors & Bioelectronics. 1991;**6**:375-379

[46] Gayet J, Haouz A, Geloso-Meyer A, Burstein C. Detection of heavy metal salts with biosensors built with an oxygen electrode coupled to various immobilized oxydases and dehydrogenases. Biosensors & Bioelectronics. 1993;**8**:177-183

[47] May LM, Russell DA. Novel determination of cadmium ions using an enzyme self—Assembled monolayer with surface plasmon resonance. Analytica Chimica Acta. 2003;**500**:119-125

[48] Malitesta C, Guascito MR. Heavy metal determination by biosensors based on enzyme immobilised by electropolymerisation. Biosensors and Bioelectronics. 2005;**20**:1643-1647

[49] Corbisier P, van der Lelie D, Borremans B, Provoost A, de Lorenzo V, Brown N, et al. Whole cell- and protein-based biosensors for the detection of bioavailable heavy metals in environmental samples. Analytica Chimica Acta. 1999;**387**:235-244

[50] Saatci E, Nistor M, Gaspar S, Csoregi E, Iscan M. Comparison of two glutathione S-transferases used in capacitive biosensors for detection of heavy metals. International Journal of Environmental Analytical Chemistry. 2007;**87**(10-11):745-754

[51] Bontidean I, Ahlqvist J, Mulchandani A, Chen W, Bae W, Mehra RK, et al. Novel synthetic phytochelatin-based capacitive biosensor for heavy metal ion detection. Biosensors and Bioelectronics. 2003;**18**:547-553

[52] Wu CM, Lin LY. Immobilization of metallothionein as a sensitive biosensor chip for a detection of metal ion by surface plasmon response. Biosensors & Bioelectronics. 2004;**20**:864-871

[53] Varriale A, Staiano M, Rossi M, D'Auria S. High-affinity binding of cadmium ions by mouse

metallothionein prompting the design of a reversed-displacement protein-based fluorescence biosensor for cadmium detection. Analytical Chemistry. 2007;**79**:5760-5762

[54] Liu Q, Cai H, Xu Y, Xiao L, Yang M, Wang P. Detection of heavy metal toxicity using cardiac cell-based biosensor. Biosensors & Bioelectronics. 2007;**22**(12):3224-3229

[55] Pohanka M. Monoclonal and polyclonal antibodies production—Preparation of potent biorecognition element. Journal of Applied Biomedicine. 2009;**7**(3):115-121

[56] Blake DA, Chakrabarti P, Khosraviani M, Hatcher FM, Westhoffi CM, Goebeli P, et al. Metal binding properties of a monoclonal antibody directed toward metal-chelate complexes. The Journal of Biological Chemistry. 1996;**271**(44):27677-27685

[57] Lin T-J, Chung M-F. Using monoclonal antibody to determine lead ions with a localized surface plasmon resonance fiber-optic biosensor. Sensors. 2008;**8**:582-593

[58] Kulkarni SJ, Joshi KS, Karve MS, Damle KA. Development of fluorescence based biosensor for estimation of heavy metal ions. In: 2011 International Conference on Bioscience, Biochemistry and Bioinformatics, Vol. 5; IPCBEE. Singapore: IACSIT Press; 2011

[59] Rodriguez BB, Bolbot JA, Tothill IE. Urease-glutamic dehydrogenase biosensor for screening heavy metals in water and soil samples. Analytical and Bioanalytical Chemistry. 2014;**380**(2):284-292

[60] Bontidean I, Berggren C, Johansson G, Csöregi E, Mattiasson B, Lloyd JR, et al. Detection of heavy metal ions at femtomolar levels using protein-based biosensors. Analytical Chemistry. 1998;**70**(19):4162-4169

[61] Oltmanns J. Biosensors for heavy metals [thesis presented for the degree of Doctor of Philosophy Edinburgh]; 2016

[62] Leth S, Maltoni S, Simkus R, Mattiasson B, Corbisier P, Klimant I, et al. Engineered bacteria based biosensors for monitoring bioavaiable heavy metals. Electroanalysis. 2002;**14**(1):35-42

Chapter 11

Principle and Development of Phage-Based Biosensors

Umer Farooq, Qiaoli Yang, Muhammad Wajid Ullah and Shenqi Wang

Abstract

Detection and identification of pathogenic bacteria is important in the field of public health, medicine, food safety, environmental monitoring and security. Worldwide, the common cause of mortality and morbidity is bacterial infection often due to misdiagnosis or delay in diagnosis. Existing bacterial detection methods rely on conventional culture or microscopic techniques and molecular methods that often time consuming, laborious and expensive, or need trained users. In recent years, biosensor remained an interesting topic for bacterial detection and many biosensors involving different bio-probes have been reported. Compared to antibodies, nucleic acids and enzymes etc., based biosensors, bacteriophages can be cheaply produced and are relatively much stable to elevated temperature, extreme pH, and diverse ionic strength. Therefore, there is an urgent need for phage-based biosensor for bacterial pathogen detection. Furthermore, bearing high affinity and specificity, bacteriophages are perfect bio-recognition probes in biosensor development for bacterial detection. In this regard, active and oriented phages immobilization is the key step toward phage-based biosensor development. This chapter compares different bacterial detection techniques, and introduces the basic of biosensor and different bio-probes involved in biosensor development. Further we highlight the involvement and importance of phages in biosensor and finally we briefed different phage immobilization approaches used in development of phage-based biosensors.

Keywords: biosensor development, bacterial detection techniques, bio-probes, phage-probe, phage immobilization

1. Introduction

The risks due to bacterial contamination and infection to healthcare system and socio-economic stability as well as to environment and food contamination have become global issues [1]. The current approaches are usually not performing well in complex mixtures of opposing microorganisms and environmental conditions devoid of enrichment step. These approaches comprise old-fashioned plating and antibodies-based assays. Therefore, in the skipping of enrichment step, almost all present experiments are not satisfactorily sensitive to sense a distinct or a very small quantity of target bacteria [2]. In contrast, the approaches like hybridization-based assays (ELISA) and polymerase chain reaction (PCR) are sensitive; however, these cannot differentiate the live cells from the dead ones, thus require an augmentation step for specificity and are laborious and expensive. These restrictions can be potentially overwhelmed by developing a biosensor. Biosensor development needs a specific and sensitive bio-probe that can withstand elevated temperature, extreme pH and remain active in diverse and complicated environment. Bacteriophages being sensitive and specific to host bacterium, and showing activity in diverse ionic concentrations are potent agents in biosensor development for detection of bacteria. Phages naturally deliver specificity in recognition of particular bacterial strain to attach, and specifically sense preferred bacterial spectra. Swift recognition offered by phage-based detection can improve the tracing and remediation of bacterial contamination [3]. The main issue that comes with development of phage based biosensor is active and oriented phages immobilization on substrate surface. The benefit of phage immobilization during biosensor development is that phages remain active for long time period, retain physiological activities with high densities, and having high bacterial cells capture efficiencies. Thus, showing improved detection limits that leads to possible development of phage-based biosensor for rapid and accurate bacterial detection [4]. Bacteriophage based biosensor development involve the following phage related approaches: (i) Observing the released phage particles during lytic cycle in the presence of host bacterium, (ii) monitoring released intracellular lysed cell component in the course of phage-mediated bacterial lysis, (iii) detection of inhibited bacterial growth in the presence of specific phages, (iv) use of stained phages for bacterial capture, and (v) observing the expression of cloned reporter gene in genetically modified phages that is expressed after bacterial infection [5].

2. Bacterial detection approaches

The conventional bacterial detection techniques such as colony count, biochemical and immunological procedures (ELISA) [6], and the modern (PCR) [7] approaches are currently widely in use; however, these approaches are time consuming as these need enrichment step. Consequently, there is a need to develop rapid and sensitive detecting methods. To this end, the use of biosensor, which can sense bacteria at diverse concentrations, are considered well applicable platform owing to their low cost, simplicity, and sensitivity [5]. **Figure 1** shows different bacterial detection approaches and **Table 1** summarizes comparative study of different bacterial detection methods.

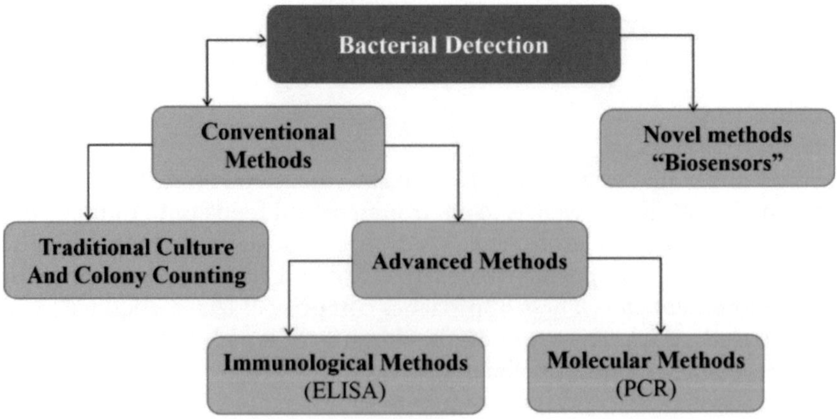

Figure 1.
Representation of various bacterial detection approaches.

Bacterial detection method	Personals	Cost and detection time	Tools	Live and dead cells detection	On spot detection	Ref.
Culture & colony count	Trained users, laborious	Cheap, 5–7 days	Simple	Yes	No	[2, 13, 14]
PCR	Trained users, laborious,	Costly, 1–4 h	Specialized	No	No	
ELISA	Trained users, laborious,	Costly, approx.: 4 h	High-tech	No	No	
Nucleic acids-based-biosensor	Simple, automatic	Expensive, 0.5–2 h	Simple	No	Yes	[2, 15, 16]
Antibodies-based-biosensor	Simple, automatic	Very expensive, 0.5–2 h	Simple	No	Yes	
Phage-based-biosensor	Simple, Automatic	Cheap, 0.5–2 h	Simple	Yes	Yes	

Table 1.
A comparison between culture and colony count, advanced molecular, and novel biosensors-based bacterial screening approaches, adapted and modified from [5].

2.1 Traditional culturing methods

In such methods, bacteria existing in a sample are cultured on different types of media so that to confirm their existence and isolate them. Two main culturing approaches are used, quantitative and qualitative. By qualitative culturing technique, the target bacterial colonies are produced on selective or differential media. In quantitative culturing technique, the specific bacteria are propagated to form specific colonies which can be calculated to evaluate the sum of microorganisms. Finally, different biochemical tests are performed [8].

2.2 Immunological methods

Immunological approaches, such as ELISA, depend upon the reaction of an antigen with a particular specific antibody. This method is unable to differentiate among living and dead cells and also antibodies production is very expensive [6].

2.3 Molecular techniques

Molecular procedures involve the use of DNA for the detection of target bacteria. For example, PCR, first pronounced in 1980s, is nowadays frequently used for detection of bacteria [7]. Molecular approaches are popular for their high sensitivity and rapidity. Dedicated apparatuses, skilled operators and expensive nature mark their rejection.

2.4 Biosensor

According to the proposed definition of biosensor by IUPAC, "Biosensor is a self-controlled imitated device, that is comprise a bio-recognition constituent (bio-prob/bio-receptor), connected to a transducer to translate the biological signal

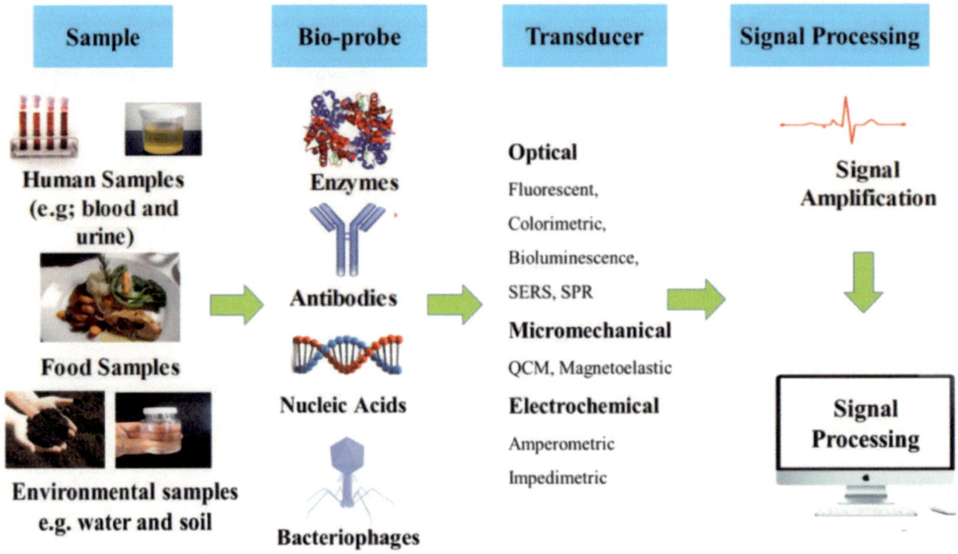

Figure 2.
Schematics representation of a generalized biosensors, reframed from [12].

into a computer readable signal and is then presented on computer and analyzed [9] (**Figure 2**). The bio-probes used in general are bacteriophage, enzyme, whole cell, nucleic acid and antibody. The transducer is electrochemical, optical, or mass based, or combination of these. Typical features of biosensors include; selectivity, reproducibility, detection limit, stability, biocompatibility, sensitivity and linearity [10]. Biosensors are commonly used in medical, diagnostic, quality control, veterinary, food and dairy industry, viral and bacterial diagnostic, agriculture industry, drug production, mining, industrial waste water control, defense and military [11]. Classification of biosensor is based on the recognition element, that is, bio-probe (bacteriophage, enzyme, whole cell, nucleic acid and antibody) used or the type of transducer (electrical, optical, or thermal signals etc.) involved. A representative biosensor is comprised of analyte (target to be sensed), bio-receptor (bio-molecule that identifies the analyte), transducer (responsible for signal transduction) and electronics (display the transduced signal) [5].

3. Bio-probes

As mentioned earlier, biosensor involves some biological recognition elements like bacteriophages [17], enzyme [18], whole cell [19], nucleic acid [20], and antibody [21], etc. These common bio-probe are briefed in the following sections:

3.1 Antibodies

To accomplish the requisite for up-to-date and fast bio-sensing schemes, antibodies (Abs) have become important affinity ligands to detect pathogens in clinical and food samples. Definitely, Abs when immobilized on a surface, these interact with specific antigens present on microbial surfaces, thus inducing a computable signal by an output detector. Abs popularity ascends from numerous benefits, for example, adaptability, ease of incorporation into diverse systems and are highly specific to their target antigens [21].

3.2 Enzymes

From the time of first biosensor (glucose sensor by Clark and Lyons in 1962), enzyme-based biosensors have shown immense progress in many applications. Enzymes are precise competent bioanalytical agents, having the ability to precisely mark out their substrates. This distinctive property mark enzymes potent implements in the development of analytical devices [18]. These biosensors company closely a biocatalyst-comprising a detecting coating with a transducer. Its operational principal is based on the catalysis and binding abilities for specific detection.

3.3 DNA/nucleic acid

The sequence of nucleic acids for a precise detection was established in 1953 and is still developing widely [20]. These biosensors involve nucleic acids as a bio-recognition-prob. The high specific binding between the two single strands of DNA (ssDNA) sequences to make double stranded DNA (dsDNA) sequence is used to develop nucleic acids-based biosensor. This technique validated to develop DNA-built-biosensor from the old-style technique like pairing of radio iso-tropic and electrophoretic separations that are costly, dangerous, and time consuming.

3.4 Cells

These biosensors involve living cells as a bio-probe and detecting component. They are constructed on the basis of living cell ability to sense the physiological parameters, and the extracellular and intracellular micro-environmental conditions, and as a result a response is produced by the reaction between cell and stimulus [19]. Microbial cells, for example fungi and bacteria are commonly used to develop whole cell based biosensors to sense particular molecules or the inclusive "condition" of the nearby environs.

3.5 Bacteriophages

Phages are virus particles, infecting and reproducing only within bacterial cells. Because of their associated evolution along with bacteria, phages have extremely specific machineries to identify and then infect their host bacteria for propagation. Phages generally have two distinctive chunks, the head comprising genetic material while the tail accountable to recognize and attach to bacterial cell [22]. Phages have several biomedical applications, and owing to their specificity they are extensively used for specific and sensitive detection of bacteria [23]. Most significant feature of phages is that they can only identify, and attack living bacterial cells. This exciting feature was well demonstrated by Fernandes et al., to detect viable, viable but not culture-able, or totally dead *Salmonella* cells on a biochip that was bio-functionalized with either phage or antibody as a bio-recognition element [24]. Interestingly dead cells were still capable to interact with the antibody, phage probes enabled a superior difference among viable and dead *Salmonella*. Additionally, some of phages are very specific and infecting only one bacterial specie [25]. This property will permit the recognition of target pathogenic bacterial species in a complex flora. Comprehensive explanation of bacteriophage based recognition elements employed as bio-probes in development of a biosensor to detect pathogenic bacteria, is outlined in the following section.

4. Bacteriophages in biosensor

A phage as a bio-recognition probe offers numerous benefits in rapid bacterial sensing [17] as they are: (•) extremely specific to their host [26], (•) ability of producing extraordinary titers of descendant phages, (•) tolerant to extreme environmental conditions like ultrahigh temperatures, organic solvents and wide-ranging pH compared to Abs, [27], (•) safe handling, (•) discriminating among dead and live bacteria as they proliferate only in live bacterial cells [28], (•) production in bulk are artless and economical. These compensations make phages as leading bio-recognition probes to develop biosensors for bacterial screening [15]. Frequently designed phage-based biosensor schemes comprise the association of whole phage or phage-constituents, infecting/capturing target bacterial cells and ultimately resulting in the production of electrical, colorimetric, fluorescent, or luminescent etc. signals, based on the available biosensing system. Hence, phages are demonstrating themselves as novel troupes for cheap, fast, sensitive and specific bacterial detection in comparison to other available platforms [29].

4.1 Reporter phage-probes

Reporter bacteriophages are genetically edited phages used to import and insert a specific gene into the genome of target bacteria. The foreign gene inserted to host bacteria is expressed, bacteria are marked based on available platforms as a colorimetric, optical, or as a fluorescent marker and thus bacterial screening is permitted [30]. Irrespective of, whether reporter bacteriophages are lysogenic or lytic, both can detect potentially the particular pathogenic bacteria. A number of genes, such as insertion of firefly *luc* or bacterial *lux* gene account for bioluminescence, β-galactosidase-*lacZ* gene, ice nucleation-*inaW* gene, and also green fluorescent protein (*GFP*) gene reported by researchers as reporter phages and detected many of Gram negative and Gram positive bacteria [31, 32].

4.2 Stained phages

Phages stained with different fluorescent dyes have been used for target bacterial detection involving various fluorescence sensing tools. Stained phage-probes can discriminate a target bacterium when they infect and attack host cells [33]. Like, phages were tagged with fluorescent quantum dots (QDs) and *E. coli* was detected at 20 colony forming units per mL in water samples within 1 h [34].

4.3 Lytic phages

Lytic phages infection results in cell burst and consequently intracellular organelles, descendant phages, and cell-lysis materials are released. Both the release of intracellular elements and released progeny phages provide a base to recognize the target bacterium [30]. For example, as a released cell component, adenosine-triphosphate can be detected through bioluminescence just after target bacterial cell lysis [35]. Also the amount of released progeny phages released after cell lysis by a particular phage is directly proportional to the amount of lysed cells and can be used for bacterial sensing [36]. The released progeny phages enumerated by various detection mechanisms such as plaque- or immuno assays, molecular methods like quantitative PCR (qPCR) and, or by isothermal nucleic acid amplification (ITNAA) [37].

4.4 Capturing phages

Phages that are immobilized on solid matrix can be utilized for capturing specific bacterial cell from contained samples. Bacteriophages have a many functionally active groups like hydroxyl group (—OH), aldehyde group(—CHO), carboxyl group(—COOH), etc., on their exteriors, giving them inimitable characteristics permitting their interaction with other materials and to interact with bacterial surface receptor molecules [38]. Consequently phages have been successfully used to capture particular bacterial cells from different samples [39, 40]. Like streptavidin actuated gold nano-particles were used to immobilize GM T4 bacteriophage particles. Delay in impedance was observed due to bacterial cells binding that marked as a sign for the existence of bacterial cells [41].

4.5 Phage receptor-binding-proteins

Some phage components display natural magnetism to host cell for example receptor-binding proteins (RBPs), but they are highly subtle to variations in environmental conditions. Phage tail bears RBPs and helps in binding to host bacterium, proceeding to insert its genetic material within the cell and cell infection is established [42]. RPBs bind to cell surface with help of specific polypeptide or polysaccharide sequences that are present on the cell surface. Poshtiban and colleagues activated magnetic beads by immobilization of RBP protein Gp047 (from phage NCTC12673). These functionally active beads were then utilized for *Campylobacter* cells withdrawal from samples of milk and broth of chicken [43].

5. Phages immobilization strategies

It's obvious from the literature that different approaches have been developed for immobilization of phages on surface of electrodes **Figure 3**. The common phage immobilization strategies on solid surfaces include physical adsorption [44], covalent bonding [45], entrapment of phages in solid matrix [46], etc.

The quantity of randomly oriented phages on solid surfaces is the most straightforward way for enhancing signal in bio-sensing scheme [47]. Deposition

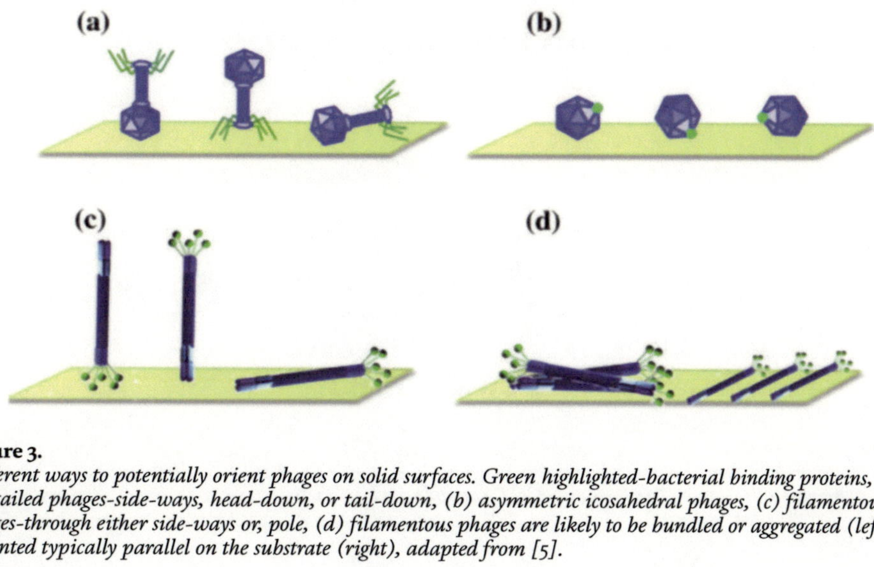

Figure 3.
Different ways to potentially orient phages on solid surfaces. Green highlighted-bacterial binding proteins, (a) tailed phages-side-ways, head-down, or tail-down, (b) asymmetric icosahedral phages, (c) filamentous phages-through either side-ways or, pole, (d) filamentous phages are likely to be bundled or aggregated (left). Oriented typically parallel on the substrate (right), adapted from [5].

of high number of phage particles creates a steric interruption between phage particles [39]. Thus number of phage particles immobilized on solid surface should not surpass a specific threshold per surface area [48]. For T4 phage, estimated optical density was 19 phages/ mm^2 area beyond that clogging was happened, resulting in reduced signal [49]. Phage particles can be simply oriented on the surface of electrode as they bear positive and negative potential on their tail fibers and head respectively. Phage immobilization strategies are briefly highlighted in the following context.

5.1 Immobilization by physisorption, electrostatic bindings and covalent bonding

Most common approach used for immobilization of phages is physisorption [50, 51]. This approach is very artless, but then again the adsorbed phage may possibly detach as of substrate surface because of shear, changes in pH, or temperature, or ionic concentrations caused in the medium that reduces principally their biosensing applications. Subsequently most phage particles having net negative charge at pH 7 [52], a number of investigators successfully used electrostatic binding for phages immobilization **Figure 4,** [52]. Also this methodology suffers due to variability and bacteriophage detachment in turn to the physico-chemical fluctuations in the analyte medium. Covalent bonding of phages offered a more stronger attachment and is not at risk to easy detachment of phages [53, 54]. Proper chemical studies can make easy selection of suitable substrate and then potential application. Covalent attachment resulted in a sophisticated bacteriophage surface mass that is principally necessary for phage application in biosensor development [55, 56]. To design bioactive surfaces with phages, phage infectivity is important or at least phage should be able to interact with host bacteria or analyte; therefore, optimization is needed to reduce the effect on bacteriophage integrity during immobilization.

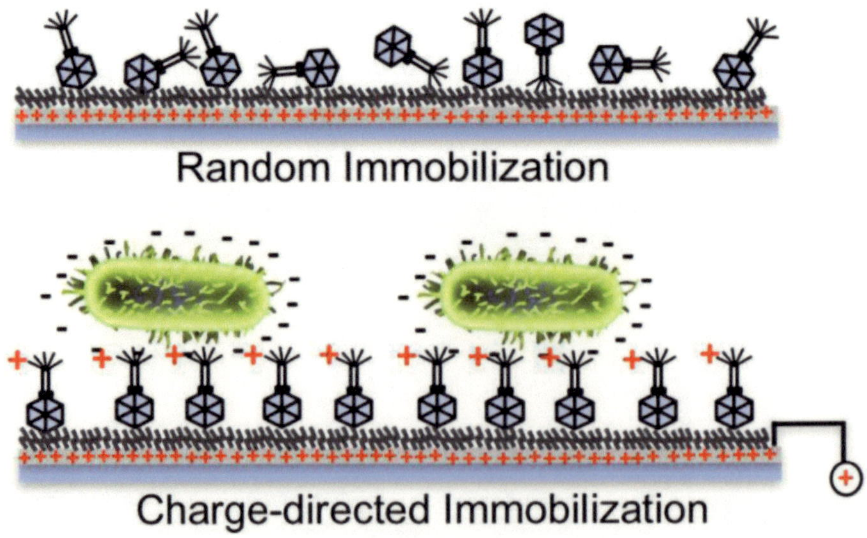

Figure 4.
Graphical representation of bacteriophage random immobilization and electrostatic, charge-directed orientated immobilization of T2 phage onto CNT electrode surface functionalized with polyethyleneimine (PEI) [57].

5.2 Phage display technologies

Bacteriophage-display tools can enable scientist to display peptides of choice present on the phage exterior, that is, phage envelope. Phages expressed peptide can consequently be adsorbed on material surfaces that are coated with peptides specific ligands **Figure 5**. Phage-display-libraries are produced by introducing DNA segments into specific phages to facilitate each phage to display a specific peptide expressed by the DNA segment inserted [58]. Technology of phage display developed as a combined influence of two central thoughts, fusion phage and combinatorial peptide libraries [59]. The first theory allows display of external peptides on bacteriophage surface [59]; while the second idea hires libraries of numerous peptides achieved in corresponding production as contrasting to production of single specific peptides [60]. Merging these two theories stemmed progress in phage-display-tools, multi-billion clone alignments of self-assembled and self-amplified bio-components [54]. It is significant to keep in mind that genetic alteration may alter the characteristics properties of bacteriophages. For example, biotin-carboxyl-carrier-protein (BCCP) gene or the cellulose-binding-module (CBM) gene to the small-outer-capsid-protein (SOCP) gene of T4 bacteriophage was attached, affecting bacteriophage infectivity, and result was decline in burst size, as well as extended latent period [61].

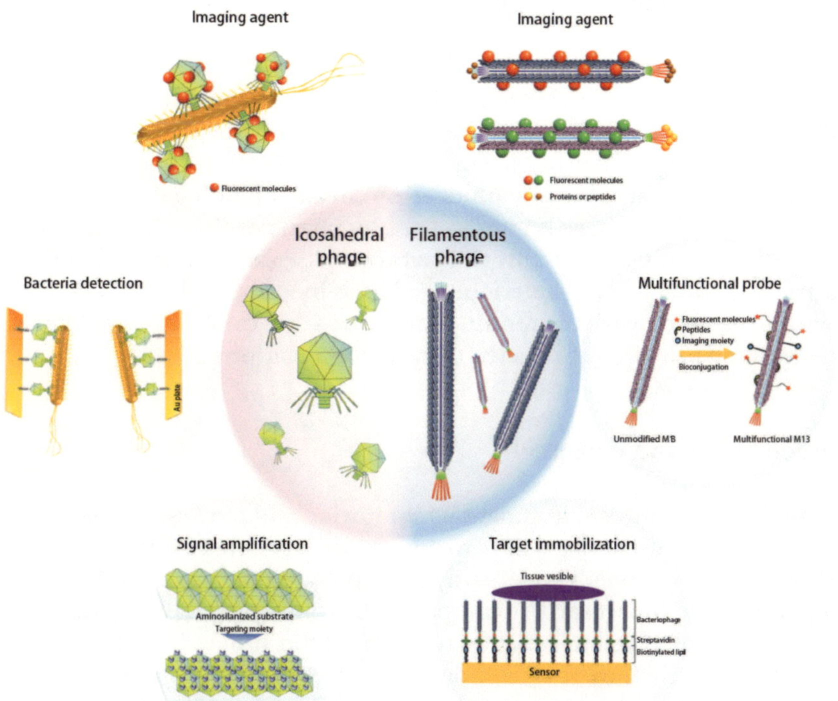

Figure 5.
Current applications of phage display technologies as imaging agents. Icosahedral phages are mostly used as, aiming on moieties for bacterial detection, and substrates for signal amplification. While filamentous phages are mostly used as multifunctional probes, and a variety of sensors [67].

5.3 Phage entrapment in porous matrix

Bacteriophages immobilization in micro-porous matrices permits them functionally and also structurally stable, keeping them active for long time period. Phages immobilization by entrapment in a porous hydrogel, (bio)polymeric agar and alginate matrices, is a tool for selection of applications where protection of phage particles essential against severe environmental conditions [62]. Additionally, entrapment might aid to maintain moisture, which is important for many phages infectivity, or keep phage particles in lyophilized condition [63]. A fruitful marketable case in point of entrapped bacteriophage in matrix is PhagoBioDerm [64] that is 0.2 mm thick, porous-polymeric-wound-dressing saturated with a mixture of biocides and lytic phages [65]. The matrices used for bacteriophage entrapment, that might possibly delay interaction of entrapped bacteriophage particles with host bacterial cells or analytes that are present in the vicinity of medium [66], marking inefficiency of phage bioactive surface.

5.4 Phage layer by layer organization

Many investigators discovered to possibly immobilize phages by alternative layering with polyelectrolytes having oppositely charges, and claimed observation of enhanced phage particle surface coverage [68, 69]. For instance, a layer by layer methodology for M13 bacteriophage was reported, and phage was sandwiched between oppositely charged layers of weak poly-electrolytes, that was capable to diffuse freely form a nearby packed phage monolayer [69].

5.5 Efficiency of immobilized phages in biosensing platforms

The effectiveness of bio-sensing approaches is mostly measured in terms of minimum limit of detection (LOD) of bacterial or other analyte. Thus researchers attempted and focused to improve the bacteriophage surface coverage for pushing detection limits. Significantly keep in mind that the LOD has not been improved biosensors where phages are immobilized by covalent binding, in comparison to the approaches where phage is immobilized by physisorption [22]. Thus, bacteriophage surface coverage is not only the factor to necessarily increase and improve the sensitivity and LOD of bacteriophage-based biosensor. Limit of detection of biosensors, based on various transduction approaches can be different depending on the working principle of selected transduction platform.

6. Conclusions and prospects

Without any doubt, environmental monitoring and food safety are the main universal worries that we humans have to oppose and are constantly struggling to take them over. In this chapter, we evidently demonstrated the principle and development phage-based biosensor. We compared the conventional phage based detection methods and briefed an introduction to different bio-probes involved in biosensors development. Further, we reviewed demonstrative phage/phage-components used in sensors development for pathogenic bacterial detection. Finally, we briefed different techniques to immobilize phages on appropriate substrate that is the major step toward phage-based biosensor development. We intend at thought-provoking and comprehensive explanations in mounting phage-based sensors and enlightening their uses for bacterial detection. By collaboration of engineers and scientists from multidisciplinary area to design a field applicable sensor and make advancements in phage-based sensors for bacterial pathogens diagnosis, we expect that this chapter might bring together the technologies related to phage-based sensors. In short, phage based biosensors in the fields of food safety, environmental monitoring and infectious disease diagnostics is vital as they are;

- Cheap (based on easy phages production)
- Highly specific
- Very sensitive
- Versatile (based on phage components)

Acknowledgements

This work was supported by National Key Research and Development Program of China under Grant 2017YFC1104402, China Postdoctoral Science Foundation (2016 M602291), the initial Research fund from CSC; and 3551 Project, Optics Valley of China.

Acronyms and abbreviations

Abs	antibodies
E. coli	*Escherichia coli*
ELISA	enzyme linked immunosorbent assay
ITNAA	isothermal nucleic acid amplification
IUPAC	International Union of Pure and Applied Chemistry
LOD	limit of detection
PCR	polymerase chain reaction
QDs	quantum dots
qPCR	quantitative polymerase chain reaction
RBPs	receptor-binding proteins
BCCP	biotin-carboxyl-carrier-protein
CBM	cellulose-binding-module
SOCP	small-outer-capsid-protein

Author details

Umer Farooq[1], Qiaoli Yang[1], Muhammad Wajid Ullah[2] and Shenqi Wang[1*]

1 Advanced Biomaterials and Tissue Engineering Centre, Huazhong University of Science and Technology, Wuhan, PR China

2 Department of Biomedical Engineering, Huazhong University of Science and Technology, Wuhan, PR China

*Address all correspondence to: shenqiwang131@hust.edu.cn

References

[1] Fleurot I, Aigle M, Fleurot R, Darrigo C, Hennekinne JA, Gruss A, et al. Following pathogen development and gene expression in a food ecosystem: The case of a *Staphylococcus aureus* isolate in cheese. Applied and Environmental Microbiology. 2014;**80**:5106-5115. DOI: 10.1128/AEM.01042-14

[2] Richter Ł, Janczuk-Richter M, Niedziółka-Jönsson J, Paczesny J, Hołyst R. Recent advances in bacteriophage-based methods for bacteria detection. Drug Discovery Today. 2018;**23**:448-455. DOI: 10.1016/J.DRUDIS.2017.11.007

[3] Lu TK, Bowers J, Koeris MS. Advancing bacteriophage-based microbial diagnostics with synthetic biology. Trends in Biotechnology. 2013;**31**:325-327. DOI: 10.1016/j.tibtech.2013.03.009

[4] Gao R, Wang Y, Tong J, Zhou P, Yang Z. Strategies for the immobilization of bacteriophages applied in the biosensors. In: Proceedings of the 2015 International Conference on Comput. Intell. Commun. Networks, CICN 2015. 2016. DOI: 10.1109/CICN.2015.42

[5] Farooq U, Yang Q, Ullah MW, Wang S. Bacterial biosensing: Recent advances in phage-based bioassays and biosensors. Biosensors & Bioelectronics. 2018. DOI: 10.1016/j.bios.2018.07.058

[6] Palumbo JD, Borucki MK, Mandrell RE, Gorski L. Serotyping of *Listeria monocytogenes* by enzyme-linked immunosorbent assay and identification of mixed-serotype cultures by colony immunoblotting. Journal of Clinical Microbiology. 2003;**41**:564-571. DOI: 10.1128/JCM.41.2.564-571.2003

[7] Uttamchandani M, Neo JL, Ong BNZ, Moochhala S. Applications of microarrays in pathogen detection and biodefence. Trends in Biotechnology. 2009;**27**:53-61. DOI: 10.1016/j.tibtech.2008.09.004

[8] Sanders SQ, Boothe DH, Frank JF, Arnoldi JW. Determination of urea with special emphasis on biosensors: A review. Journal of Food Protection. 2007. DOI: 10.4315/0362-028X-70.6.1379

[9] Qian Y, Fan T, Wang P, Zhang X, Luo J, Zhou F, et al. A novel label-free homogeneous electrochemical immunosensor based on proximity hybridization-triggered isothermal exponential amplification induced G-quadruplex formation. Sensors & Actuators, B: Chemical. 2017. DOI: 10.1016/j.snb.2017.03.152

[10] Metkar SK, Girigoswami K. Diagnostic biosensors in medicine—A review. Biocatalaysis and Agricultural Biotechnology. 2018. DOI: 10.1016/J.BCAB.2018.11.029

[11] Pundir CS, Lata S, Narwal V. Biosensors for determination of D and L- amino acids: A review. Biosensors & Bioelectronics. 2018. DOI: 10.1016/j.bios.2018.06.033

[12] Luka G, Ahmadi A, Najjaran H, Alocilja E, Derosa M, Wolthers K, et al. Microfluidics integrated biosensors: A leading technology towards lab-on-a-chip and sensing applications. Sensors (Switzerland). 2015. DOI: 10.3390/s151229783

[13] Singhal N, Kumar M, Kanaujia PK, Virdi JS. MALDI-TOF mass spectrometry: An emerging technology for microbial identification and diagnosis. Frontiers in Microbiology. 2015;**6**:1-16. DOI: 10.3389/fmicb.2015.00791

[14] Zhang L, Huang Y, Wang J, Rong Y, Lai W, Zhang J, et al. Hierarchical

flowerlike gold nanoparticles labeled immunochromatography test strip for highly sensitive detection of *Escherichia coli* O157:H7. Langmuir. 2015;**31**:5537-5544. DOI: 10.1021/acs.langmuir.5b00592

[15] Shabani A, Marquette CA, Mandeville R, Lawrence MF. Carbon microarrays for the direct impedimetric detection of *Bacillus anthracis* using gamma phages as probes. The Analyst. 2013;**138**:1434. DOI: 10.1039/c3an36830k

[16] Singh A, Poshtiban S, Evoy S. Recent advances in bacteriophage based biosensors for food-borne pathogen detection. Sensors (Switzerland). 2013;**13**:1763-1786. DOI: 10.3390/s130201763

[17] Alcaine SD, Law K, Ho S, Kinchla AJ, Sela DA, Nugen SR. Bioengineering bacteriophages to enhance the sensitivity of phage amplification-based paper fluidic detection of bacteria. Biosensors & Bioelectronics. 2016;**82**:14-19. DOI: 10.1016/j.bios.2016.03.047

[18] Liu J, Lu Y. A colorimetric lead biosensor using DNAzyme-directed assembly of gold nanoparticles. Journal of the American Chemical Society. 2003. DOI: 10.1021/ja034775u

[19] Wang J, Wu C, Hu N, Zhou J, Du L, Wang P. Microfabricated electrochemical cell-based biosensors for analysis of living cells In vitro. Biosensors. 2012. DOI: 10.3390/bios2020127

[20] Liu A, Wang K, Weng S, Lei Y, Lin L, Chen W, et al. Development of electrochemical DNA biosensors. Trends in Analytical Chemistry. 2012. DOI: 10.1016/j.trac.2012.03.008

[21] Byrne B, Stack E, Gilmartin N, O'Kennedy R. Antibody-based sensors: Principles, problems and potential for detection of pathogens and associated toxins. Sensors (Switzerland). 2009. DOI: 10.3390/s90604407

[22] Hosseinidoust Z, Olsson ALJ, Tufenkji N. Going viral: Designing bioactive surfaces with bacteriophage. Colloids Surfaces B: Biointerfaces. 2014;**124**:2-16. DOI: 10.1016/j.colsurfb.2014.05.036

[23] Schmelcher M, Loessner MJ. Application of bacteriophages for detection of foodborne pathogens. Bacteriophage. 2014;**4**:128-137. DOI: 10.4161/bact.28137

[24] Fernandes E, Martins VC, Nóbrega C, Carvalho CM, Cardoso FA, Cardoso S, et al. A bacteriophage detection tool for viability assessment of Salmonella cells. Biosensors & Bioelectronics. 2014;**52**:239-246. DOI: 10.1016/j.bios.2013.08.053

[25] Tawil N, Sacher E, Mandeville R, Meunier M. Bacteriophages: Biosensing tools for multi-drug resistant pathogens. The Analyst. 2014;**139**. DOI: 10.1039/c3an01989f

[26] Balasubramanian S, Sorokulova IB, Vodyanoy VJ, Simonian AL. Lytic phage as a specific and selective probe for detection of *Staphylococcus aureus*—A surface plasmon resonance spectroscopic study. Biosensors & Bioelectronics. 2007;**22**:948-955. DOI: 10.1016/j.bios.2006.04.003

[27] Bárdy P, Pantůček R, Benešík M, Doškař J. Genetically modified bacteriophages in applied microbiology. Journal of Applied Microbiology. 2016;**121**:618-633. DOI: 10.1111/jam.13207

[28] Altintas Z, Pocock J, Thompson KA, Tothill IE. Comparative investigations for adenovirus recognition and quantification: Plastic or natural antibodies? Biosensors & Bioelectronics. 2015;**74**:996-1004. DOI: 10.1016/j.bios.2015.07.076

[29] Wang F, Horikawa S, Hu J, Wikle HC, Chen IH, Du S, et al. Detection of *Salmonella typhimurium* on spinach using phage-based magnetoelastic biosensors. Sensors (Switzerland). 2017;**17**. DOI: 10.3390/s17020386

[30] Burnham S, Hu J, Anany H, Brovko L, Deiss F, Derda R, et al. Towards rapid on-site phage-mediated detection of generic *Escherichia coli* in water using luminescent and visual readout. Analytical and Bioanalytical Chemistry. 2014;**406**:5685-5693. DOI: 10.1007/s00216-014-7985-3

[31] Thouand G, Vachon P, Liu S, Dayre M, Griffiths MW. Optimization and validation of a simple method using P22::luxAB bacteriophage for rapid detection of *Salmonella enterica* serotypes A, B, and D in poultry samples. Journal of Food Protection. 2008;**71**:380-385. DOI: 10.4315/0362-028X-71.2.380

[32] Sharp NJ, Vandamm JP, Molineux IJ, Schofield DA. Rapid detection of *Bacillus anthracis* in complex food matrices using phage-mediated bioluminescence. Journal of Food Protection. 2015;**78**: 963-968. DOI: 10.4315/0362-028X.JFP-14-534

[33] Bhardwaj N, Bhardwaj SK, Mehta J, Kim K-H, Deep A. MOF–bacteriophage biosensor for highly sensitive and specific detection of *Staphylococcus aureus*. ACS Applied Materials & Interfaces. 2017;**9**:33589-33598. DOI: 10.1021/acsami.7b07818

[34] Yim PB, Clarke ML, McKinstry M, De Paoli Lacerda SH, Pease LF, Dobrovolskaia MA, et al. Quantitative characterization of quantum dot-labeled lambda phage for *Escherichia coli* detection. Biotechnology and Bioengineering. 2009;**104**: 1059-1067. DOI: 10.1002/bit.22488

[35] Blasco R, Murphy MJ, Sanders MF, Squirrell DJ. Specific assays for bacteria using phage mediated release of adenylate kinase. Journal of Applied Microbiology. 1998;**84**:661-666. DOI: 10.1046/j.1365-2672.1998.00393.x

[36] Cox CR, Jensen KR, Mondesire RR, Voorhees KJ. Rapid detection of *Bacillus anthracis* by γ phage amplification and lateral flow immunochromatography. Journal of Microbiological Methods. 2015;**118**:51-56. DOI: 10.1016/j.mimet.2015.08.011

[37] Brovko LY, Anany H, Griffiths MW. Bacteriophages for detection and control of bacterial pathogens in food and food-processing environment. Advances in Food and Nutrition Research. 2012;**67**:241-288. DOI: 10.1016/B978-0-12-394598-3.00006-X

[38] Ullah MW, Shi Z, Shi X, Zeng D, Li S, Yang G. Microbes as structural templates in biofabrication: Study of surface chemistry and applications. ACS Sustainable Chemistry & Engineering. 2017;**5**:11163-11175. DOI: 10.1021/acssuschemeng.7b02765

[39] Wang C, Sauvageau D, Elias A. Immobilization of active bacteriophages on polyhydroxyalkanoate surfaces. ACS Applied Materials & Interfaces. 2016;**8**:1128-1138. DOI: 10.1021/acsami.5b08664

[40] Anany H, Brovko L, El Dougdoug NK, Sohar J, Fenn H, Alasiri N, et al. Print to detect: A rapid and ultrasensitive phage-based dipstick assay for foodborne pathogens. Analytical and Bioanalytical Chemistry. 2018;**410**:1217-1230. DOI: 10.1007/s00216-017-0597-y

[41] Gervais L, Gel M, Allain B, Tolba M, Brovko L, Zourob M, et al. Immobilization of biotinylated bacteriophages on biosensor surfaces. Sensors & Actuators, B: Chemical. 2007;**125**:615-621. DOI: 10.1016/j.snb.2007.03.007

[42] Casjens SR, Molineux IJ. Short noncontractile tail machines: Adsorption and DNA delivery by podoviruses. Advances in Experimental Medicine and Biology. 2012;**726**:143-179. DOI: 10.1007/978-1-4614-0980-9_7

[43] Poshtiban S, Javed MA, Arutyunov D, Singh A, Banting G, Szymanski CM, et al. Phage receptor binding protein-based magnetic enrichment method as an aid for real time PCR detection of foodborne bacteria. The Analyst. 2013;**138**:5619. DOI: 10.1039/c3an01100c

[44] Mejri MB, Baccar H, Baldrich E, Del Campo FJ, Helali S, Ktari T, et al. Impedance biosensing using phages for bacteria detection: Generation of dual signals as the clue for in-chip assay confirmation. Biosensors & Bioelectronics. 2010;**26**:1261-1267. DOI: 10.1016/j.bios.2010.06.054

[45] Shabani A, Zourob M, Allain B, Marquette CA, Lawrence MF, Mandeville R. Bacteriophage-modified microarrays for the direct impedimetric detection of bacteria. Analytical Chemistry. 2008;**80**:9475-9482. DOI: 10.1021/ac801607w

[46] Arter JA, Taggart DK, McIntire TM, Penner RM, Weiss GA. Virus-PEDOT nanowires for biosensing. Nano Letters. 2010;**10**:4858-4862. DOI: 10.1021/nl1025826

[47] Singh A, Glass N, Tolba M, Brovko L, Griffiths M, Evoy S. Immobilization of bacteriophages on gold surfaces for the specific capture of pathogens. Biosensors & Bioelectronics. 2009;**24**:3645-3651. DOI: 10.1016/j.bios.2009.05.028

[48] Naidoo R, Singh A, Arya SK, Beadle B, Glass N, Tanha J, et al. Surface-immobilization of chromatographically purified bacteriophages for the optimized capture of bacteria. Bacteriophage. 2012;**2**:15-24. DOI: 10.4161/bact.19079

[49] Olsson ALJ, Wargenau A, Tufenkji N. Optimizing bacteriophage surface densities for bacterial capture and sensing in quartz crystal microbalance with dissipation monitoring. ACS Applied Materials & Interfaces. 2016;**8**:13698-13706. DOI: 10.1021/acsami.6b02227

[50] Huang S, Yang H, Lakshmanan RS, Johnson ML, Chen I, Wan J, et al. The effect of salt and phage concentrations on the binding sensitivity of magnetoelastic biosensors for bacillus anthracis detection. Biotechnology and Bioengineering. 2008. DOI: 10.1002/bit.21995

[51] Park MK, Weerakoon KA, Oh JH, Chin BA. The analytical comparison of phage-based magnetoelastic biosensor with TaqMan-based quantitative PCR method to detect Salmonella typhimurium on cantaloupes. Food Control. 2013. DOI: 10.1016/j.foodcont.2013.02.026

[52] Anany H, Chen W, Pelton R, Griffiths MW. Biocontrol of listeria monocytogenes and *Escherichia coli* O157:H7 in meat by using phages immobilized on modified cellulose membranes. Applied and Environmental Microbiology. 2011;**77**:6379-6387. DOI: 10.1128/AEM.05493-11

[53] Tawil N, Sacher E, Mandeville R, Meunier M. Strategies for the immobilization of bacteriophages on gold surfaces monitored by surface plasmon resonance and surface morphology. Journal of Physical Chemistry C. 2013. DOI: 10.1021/jp400565m

[54] Tlili C, Sokullu E, Safavieh M, Tolba M, Ahmed MU, Zourob M. Bacteria screening, viability, and confirmation assays using bacteriophage-impedimetric/loop-mediated isothermal amplification dual-response biosensors. Analytical Chemistry. 2013;**85**:4893-4901. DOI: 10.1021/ac302699x

[55] He F, Xiang M, Mi X. A new bacteriophage-modified piezoelectric sensor for rapid and specific detection of mycobacterium. Analytical Letters. 2012;**45**:1242-1253. DOI: 10.1080/00032719.2012.673106

[56] Karoonuthaisiri N, Charlermroj R, Morton MJ, Oplatowska-Stachowiak M, Grant IR, Elliott CT. Development of a M13 bacteriophage-based SPR detection using Salmonella as a case study. Sensors & Actuators, B: Chemical. 2014;**190**:214-220. DOI: 10.1016/j.snb.2013.08.068

[57] Zhou Y, Marar A, Kner P, Ramasamy RP. Charge-directed immobilization of bacteriophage on nanostructured electrode for whole-cell electrochemical biosensors. Analytical Chemistry. 2017;**89**:5734-5741. DOI: 10.1021/acs.analchem.6b03751

[58] Petrenko VA, Jayanna PK. Phage protein-targeted cancer nanomedicines. FEBS Letters. 2014. DOI: 10.1016/j.febslet.2013.11.011

[59] Smith GP, Petrenko VA. Phage display. Chemical Reviews. 1997. DOI: 10.1021/cr960065d

[60] Geysen HM, Schoenen F, Wagner D, Wagner R. Combinatorial compound libraries for drug discovery: An ongoing challenge. Nature Reviews. Drug Discovery. 2003. DOI: 10.1038/nrd1035

[61] Tolba M, Minikh O, Brovko LY, Evoy S, Griffiths MW. Oriented immobilization of bacteriophages for biosensor applications. Applied and Environmental Microbiology. 2010;**76**:528-535. DOI: 10.1128/AEM.02294-09

[62] Ma Y, Pacan JC, Wang Q, Sabour PM, Huang X, Xu Y. Enhanced alginate microspheres as means of oral delivery of bacteriophage for reducing *Staphylococcus aureus* intestinal carriage. Food Hydrocolloids. 2012. DOI: 10.1016/j.foodhyd.2010.11.017

[63] Alfadhel M, Puapermpoonsiri U, Ford SJ, McInnes FJ, Van Der Walle CF. Lyophilized inserts for nasal administration harboring bacteriophage selective for *Staphylococcus aureus*: In vitro evaluation. International Journal of Pharmaceutics. 2011. DOI: 10.1016/j.ijpharm.2011.07.006

[64] Jikia D, Chkhaidze N, Imedashvili E, Mgaloblishvili I, Tsitlanadze G, Katsarava R, et al. The use of a novel biodegradable preparation capable of the sustained release of bacteriophages and ciprofloxacin, in the complex treatment of multidrug-resistant *Staphylococcus aureus*-infected local radiation injuries caused by exposure to Sr90. Clinical and Experimental Dermatology. 2005. DOI: 10.1111/j.1365-2230.2004.01600.x

[65] Katsarava R, Beridze V, Arabuli N, Kharadze D, Chu CC, Won CY. Amino acid-based bioanalogous polymers. Synthesis, and study of regular poly(ester amide)s based on bis(a-amino acid) α,ω-alkylene diesters, and aliphatic dicarboxylic acids. Journal of Polymer Science, Part A: Polymer Chemistry. 1999. DOI: 10.1002/(SICI)1099-0518(19990215)37:4<391::AID-POLA3>3.0.CO;2-E

[66] Balcão VM, Moreira AR, Moutinho CG, Chaud MV, Tubino M, Vila MMDC. Structural and functional stabilization of phage particles in carbohydrate matrices for bacterial biosensing. Enzyme and Microbial Technology. 2013. DOI: 10.1016/j.enzmictec.2013.03.001

[67] Lee JW, Song J, Hwang MP, Lee KH. Nanoscale bacteriophage biosensors beyond phage display. International Journal of Nanomedicine. 2013. DOI: 10.2147/IJN.S51894

[68] Lin Y, Su Z, Niu Z, Li S, Kaur G, Lee LA, et al. Layer-by-layer assembly of viral capsid for cell adhesion. Acta Biomaterialia. 2008. DOI: 10.1016/j.actbio.2008.02.026

[69] Yoo PJ, Nam KT, Qi J, Lee S-K, Park J, Belcher AM, et al. Spontaneous assembly of viruses on multilayered polymer surfaces. Nature Materials. 2006. DOI: 10.1038/nmat1596

Chapter 12

Applications of Phage-Based Biosensors in the Diagnosis of Infectious Diseases, Food Safety, and Environmental Monitoring

Umer Farooq, Muhammad Wajid Ullah, Qiaoli Yang and Shenqi Wang

Abstract

Environmental pollution and food safety are becoming serious concerns to human health in developing countries. To combat such issues, researchers have developed different approaches for on-spot detection and screening of infectious disease, caused by pathogens and toxins in food and water samples. One such approach is the development of phage- and phage-component-based sensors that are highly specific, sensitive, rapid, efficient, cheap, and portable analyte screening platforms. Such sensors overcome the limitations of conventional screening approaches. This chapter highlights different food and environmental contaminations and represents the potential of phage-based biosensor for bacterial detection. It summarizes different applications of phage-based sensors in the fields of food safety and environmental monitoring and highlights current challenges and perspective. In general, this chapter brings together the technologies related to phage-based sensors and food and environmental safety, by compiling the efforts of engineers and scientists from multidisciplinary areas.

Keywords: bacteriophage, biosensor, infectious diseases, food safety, environmental monitoring

1. Introduction

Many pathogenic bacteria like *Streptococcus*, *Mycobacterium*, *Pseudomonas*, *Salmonella*, *Shigella*, etc. are causing different diseases in humans, resulting in several outbreaks and epidemics of diseases worldwide. Every year, millions of individuals get infected by these bacteria, while the common sources of infections are clinical, food-borne, airborne, and/or waterborne [1]. Clinical, food, and environmental contaminations are the eternal challenges worldwide in the healthcare systems and food safety and environmental monitoring. Irrespective of comprehensive struggles to fight such pathogenic bacteria, the numbers of clinical, food, and environment-related diseases are increasing every year [2]. As a solution to the problem, the development of biosensors especially phage-based biosensors for bacterial detection in clinical, food, and environmental samples has remained a hot topic since the last few decades. Phages in biosensor proved themselves as unique bio-probes, owing to their selectivity, specificity, and withstanding harsh environmental conditions. Establishing phage-based biosensors for application in food safety and environmental monitoring is a motivating and interesting research topic and is the urgent need of this modern era. The key point is to enhance phage-based cheap recognition tools with maximum levels of selectivity, consistency, and sensitivity with minimum times of assay. Significant struggles have been dedicated on enhancing the transducer surface of biosensor for improved detection and sensitivity. Phage-based bio-probes have been used in transducer development for several analytical approaches to offer specific and selective detection. Bacteriophages as a bio-probe have been successfully applied for bacterial detection in clinical samples (urine) [3], food samples (milk, tomatoes) [4], and environmental samples (river water) [5]. Furthermore, different analytical approaches relying on phage-based bio-probes have been reported like electrochemical [6], bioluminescence [7], fluorescence [8], mass spectrometry [9], magnetoelastic [4], surface plasmon resonance [10], lateral flow assay [11], etc. In the following context, we will review biosensor transduction platforms involving phage-based probes for transducer development to detect infectious bacteria in the field of food and environmental safety monitoring [12]. In this chapter we will highlight applications of different phage-based analytical approaches for bacterial detection in clinical, food, and environmental samples.

2. Phage-based biosensors for infectious pathogen detection

Bacteriophage as a bio-probe has been used in different transduction platforms for detection of pathogenic bacteria, which are briefed as follows:

2.1 Phage optical biosensors

Optical phage-based sensors owing from their reasonably rapid screening, sensitivity, and flexibility to a broad-ranging assay situations have been extensively explored for bacterial detection. Optical methods are classified into two core subclasses on the basis of their working principles, label-free and labeled. The best frequently used optical methods for bacterial screening are fluorescence spectrometry [8], surface plasmon resonance (SPR) [10], and bio- or chemiluminescence [13]. In the subsequent subsection, our focus is on phage bio-probe-based optical biosensors for detection of pathogens with special emphasis on food safety and environmental monitoring. **Figure 1** represents a reporter phage-based optical sensing scheme.

2.1.1 Phage-SPR-based sensors

Surface plasmon resonance (SPR) works on the principle of oscillation phenomenon that happens between the interfaces of any two materials. The change in the refractive index close to the sensor surface caused by contact of target analyte in the medium with the bio-probe (phage) present on transducer surface is measured by SPR biosensors. Phages have been widely immobilized as bio-probes on the surfaces of SPR transducers to offer facility of specific recognition of bacterial detection. The immobilized phage on SPR transducer successfully detected *E. coli* K12 [15], *S. aureus* [16], methicillin-resistant *Staphylococcus aureus* (MRSA), and *E. coli* O157:H7 [17]. Typically, the LOD was ranging from 10^2 to 10^3 CFU/mL. Phage RBPs have been utilized as bio-probes in SPR approaches for specific bacterial screening, such as Singh et al.'s activated gold-coated plates, by immobilizing genetically engineered tailspike proteins from P22 phage to demonstrate selective, specific, and real-time *Salmonella* detection with 10^3 CFU/mL sensitivity [18].

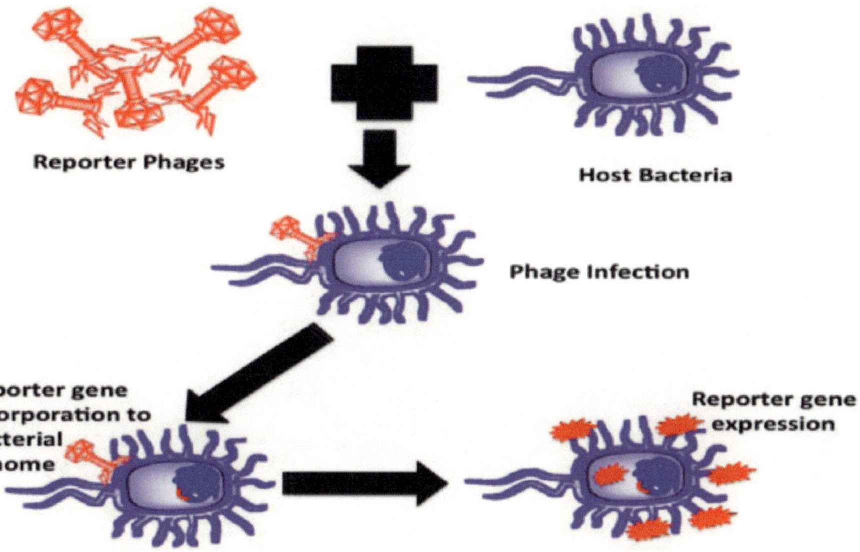

Figure 1.
A graphical representation of target pathogen detections based on reporter phage, adapted from [14].

2.1.2 Phage-bioluminescence sensors

For bacterial quantitative detection in samples, bioluminescence analyses that are rapid, sensitive, and simple are used by assessing the emitted light from intracellular components. Bacterial lysis is the first step of this type assays, to discharge intracellular cell components followed by reaction with luciferase and are screened by bioluminescent. A lytic bacteriophage is involved as a bio-recognition probe for target bacterial detection following lysis. Infectious bacteria like *E. coli* and *Salmonella Newport* were detected by an adenosine triphosphate bioluminescence assay using lytic bacteriophage as bio-probe lysis of target bacterial cell [19]. The sensitivity was enhanced 10–100-folds by addition of adenylate kinase as an alternate cell marker, while less than 10^4 CFU/mL of *E. coli* was reported in <1 h [19]. Later it was demonstrated that the quantity of discharged adenylate kinase from lysed cells is dependent on the growth stage, bacterial type, the infection time, and the phage type [20].

2.1.3 Phage-SERS-based sensors

An innovative Raman method, i.e., surface-enhanced Raman spectroscopy (SERS), is enhancing the intensity by vibrational absorbance of definite elements when they are near the surface of nano-organized noble metals by the influence of numerous orders of magnitude. The improved intensity of SERS method is dependent on the molecules' capability to release a Raman signal and the contained fields of plasmon in their neighborhood [21]. For instance, a report stated a phage-SERS biosensor for *E. coli* detection using phage immobilization on nano-figured thin sheet of silver over substrates of silica (**Figure 2**) [22] established by exploitation of metallic nanosculptured thin silver film. The silver film exterior is activated by self-assembled monolayer of 4-aminothiophenol and glutaraldehyde for T4 immobilization to screen *E. coli*. As a reporter molecule, 4-aminothiophenol monitored the Raman band enhancement. Other reports of phage-SERS-based biosensors have been reported and are briefed in **Table 1**.

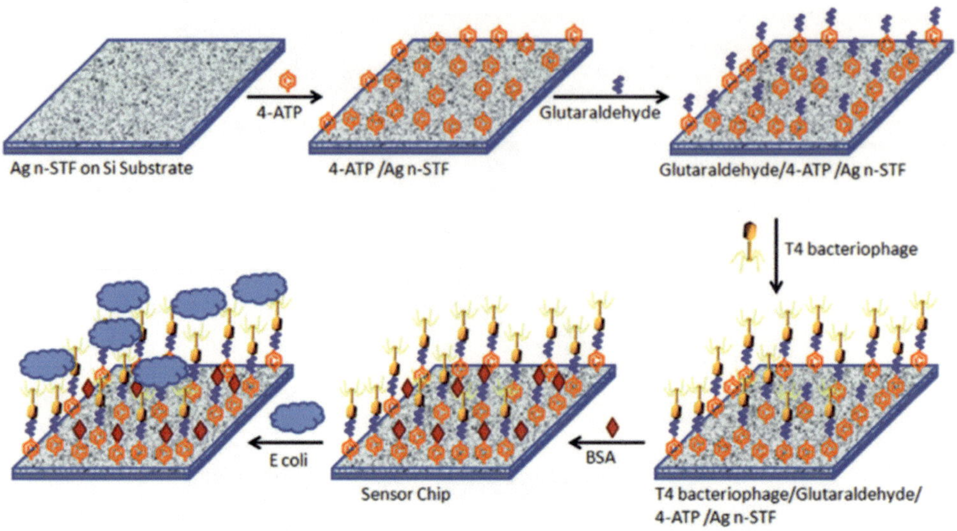

Figure 2.
Schematic representation of phage-SERS-based sensor. Adapted from [22].

Transducer	Phage-based bio-probe	Target bacteria/ analyte	Sample	Detection limit	Ref.
SPR	T4 phage	*E. coli K12*	PBS	7×10^2 CFU/mL	[15]
	T4 phage	*E. coli O157*:H7	PBS	10^3 CFU/mL	[17]
	BP14 phage	MRSA	PBS	10^3 CFU/mL	[17]
	scFv phages	*L. monocytogenes*	—	2.10^6 CFU/mL	[53]
	12,600 phage	*S. aureus*	—	10^4 CFU/mL	[16]
Luminescence	*lacZ* T4 phage	*E. coli* B	Water	10 CFU/mL	[7]
	SJ2 phage	*S. enteritidis*	—	10^3 CFU/mL	[20]
	Pap1 phage	*P. aeruginosa*	Milk, urine	56 CFU/mL	[3]
	Lytic phage	*Listeria innocua*	—	$>10^4$ CFU/mL	[54]
	Shfl25875	*S. flexneri*	Stool	10^3 CFU/g	[55]
LFA	B4 phage	*B. cereus*	Buffer	1×10^4 CFU/mL	[11]
	Gamma phage	*B. anthracis*	—	2.5×10^4 CFU/mL	[56]
	T7 phage	*E. coli*	Broth	10^3 CFU/mL	[47]
Fluorescent	P22 phage	*S. typhimurium*	Milk	1 CFU/24 mL	[57]
	P-*S. aureus*-9	*S. aureus*	PBS	2.47×10^3 CFU/L	[58]
	Wβ phage	*B. anthracis*	Soil	10^4 CFU/g	[50]
	O157-IOV 4	*E. coli O157*:H7	Milk	4.9×10^4 CFU/mL	[59]
	PP01 phage	*E. coli* O157:H7	Apple juice	1 CFU/mL	[60]
	PDPs	TNT	—	10 μg/mL	[61]
	T7 phage	*E. coli*	LB broth	10 CFU/mL	[62]
QCM	Filamentous phage	*S. typhimurium*	—	10^2 CFU/mL	[35]
	Wild-type	*E. coli* K12	—	10^3 CFU/mL	[6]
	T4 phage	*E. coli*	Milk	Few CFU/mL	[9]

Transducer	Phage-based bio-probe	Target bacteria/ analyte	Sample	Detection limit	Ref.
SERS	T4 phage	*E. coli B*	Buffer	150 CFU/mL	[22]
	Phage 12,600	MRSA	—	—	[63]
	P9b phage	*P. aeruginosa*	Clinical samples	10^3 CFU/mL	[64]
	A511 phage	*L. monocytogenes*	—	6.1×10^7pfu/mL	[65]
Magnetoelastic	E2 phage	*S. typhimurium*	—	5×10^2 CFU/mL	[66]
	JRB7 phage	*B. anthracis*	—	Spores	[67]
	E2 phage	*S. typhimurium*	Romaine lettuce	5×10^2 CFU/mL	[68]
	Phage	*S. typhimurium*	—	1.5×10^3 CFU/mm^2	[69]
Amperometric	B1-7064 phage	*B. cereus*	—	10 CFU/mL	[70]
	M13 phage	*E. coli TG1*	—	1 CFU/mL	[71]
Impedimetric	T4 phage	*E. coli*	—	10^4 CFU/mL	[72]
	T2 phage	*E. coli* B	Broth	10^3 CFU/mL	[73]
	Lytic phage	*S. Newport*	—	10^3 CFU/mL	[74]
	Gamma phage	*B. anthracis Str*	Water	10^3 CFU/mL	[75]
	T4 phage	*E. coli* B	Water, milk	800 CFU/mL 100 CFU/mL	[76]
	Endolysin Ply500	*L. monocytogenes*	Milk	10^5 CFU/mL	[77]

SPR, surface plasmon resonance; scFv, single-chain variable fragment; MRSA, methicillin-resistant Staphylococcus aureus; PBS, phosphate-buffered saline; TNB, trinitrobenzene; TNT, trinitrotoluene; QCM, quartz crystal microbalance; QD, quantum dot; SERS, surface-enhanced Raman spectroscopy; LFA, lateral flow assay; HRP, horseradish peroxidase; CFU, colony-forming unit; PFU, plaque-forming unit; E. coli, Escherichia coli; S. arlettae, Staphylococcus arlettae; B. anthracis, Bacillus anthracis; P. aeruginosa, Pseudomonas aeruginosa; S. flexneri, Shigella flexneri; S. Newport, Salmonella Newport; S. typhimurium, Salmonella typhimurium; S. aureus, Staphylococcus aureus; LB, Luria-Bertani broth.

Table 1.
Applications of phage/phage components in detection of infectious pathogen and other deadly analytes related to food safety and environmental monitoring, where transduction platform used, target analyte/bacteria, sample processed, and limit of detection are briefed with reported literature.

2.1.4 Phage-fluorescent sensor

In fluorescent-phage-based sensor techniques, fluorescently stained phages are utilized as marking agents for the detection of bacterial cells. Fluorescently labeled phages are identified followed by binding to specific host bacterial cell. The composite of bacteriophage-bacteria is then sensed by means of flow cytometry or epi-fluorescent filter approach. A combination of immunomagnetic separation with fluorescent method is detected between 10 and 10^2 CFU/mL of pathogenic bacteria *E. coli* O157:H7 after 10 h augmentation in artificially contaminated milk [23] and 10^4 CFU/mL in sample of broth medium [24]. Additional improvement in the sensitivity of this method was reported by using fluorescent quantum dots (QDs) for phage labeling [25]. Also fluorescent-based sensors have been used for bacterial toxin recognition. Phage display was applied to choose a peptide (12-mer) that was able to attach to *staphylococcal enterotoxin B* (SEB) that is responsible for food poisoning [26]. This approach permitted toxin sensing and detected 1.4 ng of SEB/sample well with the help of fluorescence immunoassay and involved a fluorescently stained SEB binding bacteriophage. Also array-based sensors have been established following the same principle for simultaneous detection of

Bacillus globigii, MS2 bacteriophage, and also SEB [27]. The typically reported sensitivity until now is about 20 CFU/mL by epi-fluorescent microscopic platform [25] and is 1 CFU/mL by flow cytometric recognition approach [28].

2.1.5 Phage-colorimetric sensors

Sensing based on changes in color allows the use of simple diagnostic systems like spectrophotometers, or even involving smartphones, and both of them are comparatively common and feasible. Designed colorimetric phage-based biosensors are mostly based and integrated on the utilization of reporter bacteriophages that carry genes coding for reporter enzymes. The foremost colorimetric sensor based on phage was to detect *Salmonella* ice nucleation sensor using reporter gene *inaW* [29]. Expression of ice nucleation protein was induced upon infection, interrupting the cell, and was consequently observed by the addition of an indicator dye (orange colored) [30]. Other serviceable reporter genes that have been successfully used with various colorimetric substrates are *celB* and *lacZ* segments encoding β-galactosidase and β-glycosidase [31]. More recently enhanced phage-based colorimetric technique has been reported to be integrating and coupling with novel technologies like surface plasmon [32], macroscope and smartphone [33], and lateral flow assay [11]. Other colorimetric phage-based biosensors established in recent years are briefed in **Table 1**.

2.2 Phage-based micromechanical sensors

Representative micromechanical biosensor (magnetoelastic) is expressed in **Figure 3**, involving E2 phage for detection of *S. typhimurium* on tomato and spinach leaves. Further micromechanical-phage-based biosensors are briefed in the following context.

2.2.1 Phage-QCM-based sensors

Quartz crystal microbalance (QCM) sensors are mass-based sensors that are highly sensitive with the ability of detecting nanogram variations in mass. QCM

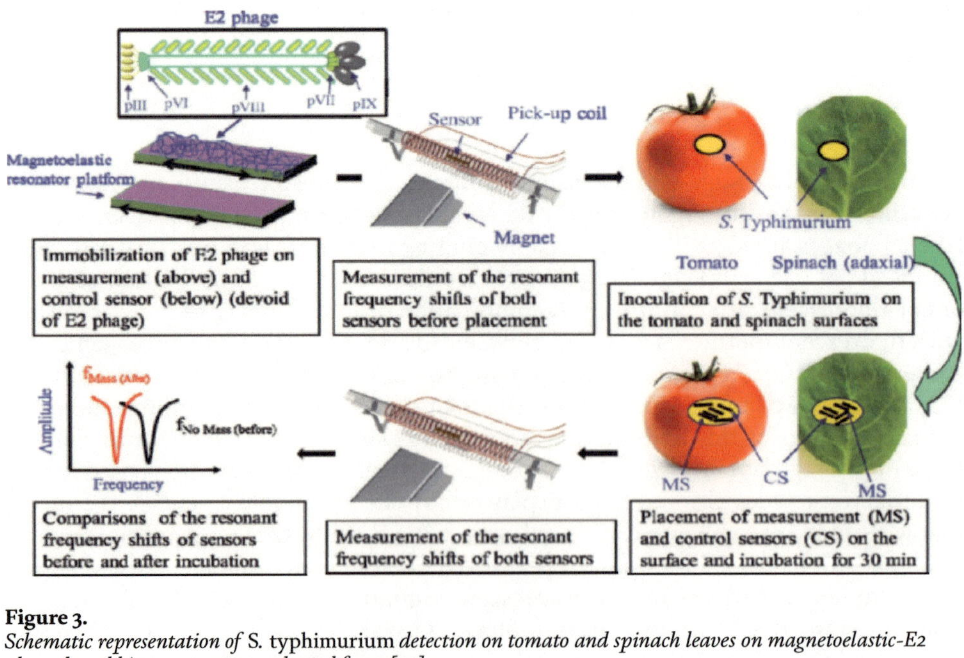

Figure 3.
Schematic representation of S. typhimurium *detection on tomato and spinach leaves on magnetoelastic-E2 phage-based biosensor system, adapted from [34].*

biosensors are functionalized by a very thin piezoelectric film having both sides coated with two conductive electrodes. Mechanical resonance is stimulated by electrical field application through the quartz crystal.

Consequently, QCM-based biosensors could be established to quantify the mass of many target analytes by immobilization of individual bio-probes on the surface of sensor. Phages as bio-probes can be conjugated with QCM biosensors for selective screening of bacterial cells. For instance, physically adsorbed bacteriophages around 3×10^{10} PFU/cm^{-2} on the surface of piezoelectric transducer provided a very rapid and sensitive platform for *Salmonella typhimurium* detection. This immobilized bacteriophage on QCM biosensor had a LOD of 10^2 CFU/mL having a broad linear range of 10^0–10^7 CFU/mL and a quick reaction and detection time of less than 3 min [35]. Other reports of phage-based QCM sensor applications in detection of infectious bacteria in food safety and environmental monitoring are briefed in **Table 1**.

2.2.2 Phage magnetoelastic sensors

Magnetoelastic sensors are prepared from materials having magnetoelastic property, i.e., magnetism and elasticity, and they contract/extend on excitation by alternative-current-magnetic field. The resonance frequency depends on the viscosity/mass adjacent to the surface of the resonating material. Magnetoelastic devices are used for detection of biological and chemical analytes by integration of bio-probes like phages on the biosensor surface and might be functional in gaseous, static, liquid, or flowing condition [21]. Likewise, E2 bacteriophage was genetically modified for specific detection of *S. typhimurium* in samples of food [36], on spinach leaves [37], and in apple juice, tomato, or milk [38], and all these magnetoelastic biosensors displayed outstanding selectivity and specificity. In addition, E2 bacteriophage-based magnetoelastic biosensors expressed tremendous stability when exposed to severe environmental conditions [39]. ME-lytic phage-based biosensor was reported to detect MRSA bacteria. In the evaluation based on varied immobilization times (10, 30, 90, 270, 810, and 2430 min) and bacteriophage concentrations (10^8–10^{12} PFU/mL), lytic phage binding to ME sensor surface was established for optimal conditions. The optimal immobilization time and concentration in PFU/mL for effective binding of phage to ME sensor surface was calculated as 30 min and 10^{11} PFU/mL, respectively. This ME-based biosensor approach was used successfully for detection of MRSA bacteria with LOD of 10^3 CFU/mL [40].

2.3 Phage-based electrochemical biosensors

A schematic representation of electrochemical biosensor of nanoflowers—AuNPs and Thi-phage composite—for *E. coli* detection is illustrated in **Figure 4**.

2.3.1 Phage-amperometric biosensors

Among the electrochemical detection methods, amperometry has been most commonly used for detection of pathogenic bacteria and offered an improved sensitivity platform related to other electrochemical approaches. Electrochemical amperometric biosensor involves a working electrode (having bio-probe) and a reference electrode. For current production in the analyte sample, a bias potential is passed on these electrodes. The produced current is directly dependent on the degree of electron transfer that fluctuates with changes in analyte's ionic concentration. Simply, amperometric sensors detect ionic changes in the solution by determining the variations in electric current. Several approaches have been established for detection of foodborne pathogenic bacteria based on phage-amperometric

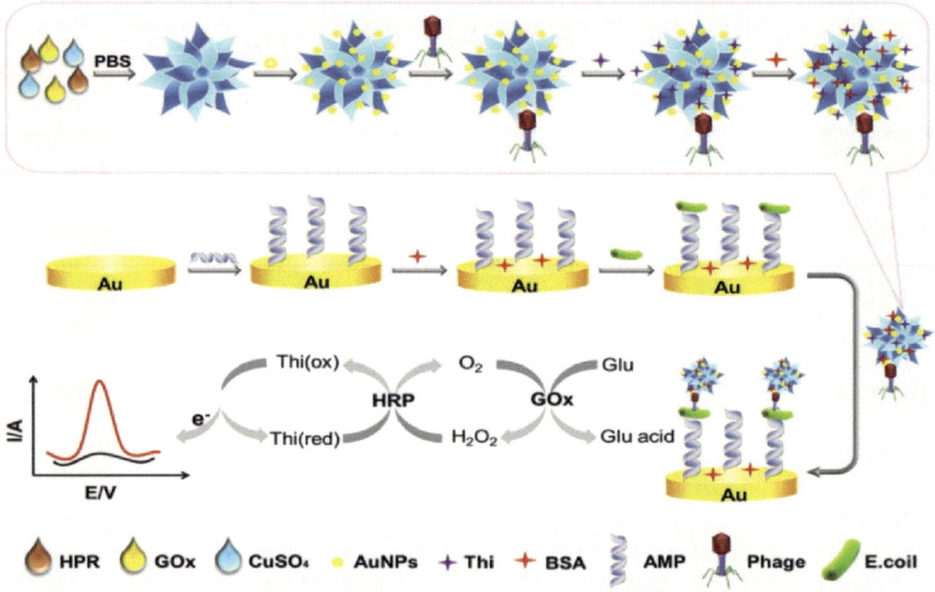

Figure 4.
Illustration of the establishment of composites (nanoflowers—AuNPs and Thi-phage) and E. coli *electrochemical screening, adapted from [41].*

biosensors. Amperometric method integrated with bacteriophage typing was reported to specifically detect bacteria like *E. coli* K12, *Bacillus cereus, and Mycobacterium smegmatis* [42]. The working principle of this biosensor was phage infection that resulted in bacterial cell lysis, subsequently releasing bacterial cell contents, like enzymes and other cell debris, into the test sample. This enzymatic release can be sensed and measured involving particular substrate. The reaction product is oxidized or reduced at working and reference electrodes, resulting in current generation [43].

2.3.2 Phage impedimetric sensors

Electrochemical impedance spectroscopy (EIS)-based sensors determine the fluctuations in impedance as a result of interactions between bio-probe and the analyte. EIS-based sensors have been utilized for bacterial detection by observing the variations on interface of solution-electrode because of the microbial capture on the biosensor surface. The target analyte binding on the sensor surface typically raises the impedance because of the insulating behavior. Phages have been utilized as a sandwiched cross-linker between bacterial cell and the electrode surface. An effective phage-EIS-based platform was reported for recognition of *E. coli* bacterial cells by T4 bacteriophage immobilization on the surface of activated carbon screen-printed electrode with LOD of ~10^4 CFU/mL [44]. By increasing bacterial concentration, a decrease in impedance was observed, which was differing from ordinary binding of intact bacterial cells on EIS biosensor. The motive behind this type of observations was because of the lytic activity of bacteriophages that directed cell lysis and the release of ionic cellular contents and alternatively a rise in conductivity. The detection was specific, and they confirmed the specificity by using *Salmonella* as a negative control. Other reports of impedimetric phage-based detections are summarized in **Table 1**.

3. Phage-based biosensors in food safety and environmental monitoring

Bacteriophage-based biosensors have been established for broad range of applications in food and environmental contaminant detection, for example, pathogens, toxins, and other environmental pollutants. Pathogens causing food contaminations are the supreme common objects of bacteriophage-based biosensors. One more field wherever bacteriophages are utilized as bio-recognition probes is clinical diagnostics of infectious diseases as explained in Section 2. **Table 1** sums up various whole phage/phage component-based biosensor applications in food safety, environmental monitoring, and infectious disease diagnosis. As this chapter does not cover all the reported methodical explanations and applications, therefore interested bibliophiles are referred to the latest literature. For potential future on-site applications, few of the most recent phage-based biosensors for pathogen detection in food and water are briefed as follow.

3.1 Food safety

Magnetoelastic (ME) phage-based biosensor was compared with TaqMan-based qPCR for *Salmonella typhimurium* detection on cantaloupe surface. LOD of both approaches was calculated by successive inoculation of cantaloupe surfaces with *S. typhimurium* suspensions. LOD of *S. typhimurium* was 2.47 ± 0.50 log CFU/2 mm^2 and 1.35 ± 0.07 log CFU/2 mm^2 area of cantaloupe surface and 6.28 and 2.41% by ME phage-based biosensor and qPCR, respectively. This comparison revealed that phage-based ME biosensor is more encouraging and an on-site applicable method to detect *S. typhimurium* on fresh fruit and vegetable surfaces [4]. In another report that was based on fluorescence imaging, *Salmonella* detection was reported involving bacteriophage-derived peptides that bind to *Salmonella enterica* (serotype Typhimurium) cells. In this report, ME biosensor coated with C4–22 phage was used to evaluate and detect *Salmonella* in/on chicken meat. In the case of on-surface detection approach, phage C4–22-based biosensor confirmed *Salmonella* binding capacity 12 times higher than control with no-phage-based sensor, while *Salmonella* cells at concentration of 7.86×10^5 CFU spiked per mm^2 area. In the case of in-chicken meat approach, phage C4–22 biosensors were inserted at varied depths below the surface of chicken meat (0.1, 0.5, 1.0 cm) after inoculation of *Salmonella* on the surface. The latter approach presented 23.27–33% of *Salmonella* cell absorption up to 0.1-cm deep under the surface [45].

P. aeruginosa was detected by lytic phage PaP1 displaying high specificity. For label-free *P. aeruginosa* detection, ECL biosensor involving PaP1 was developed. Biosensor was fabricated on glass carbon electrode surface through deposition of PaP1-conjugated carboxyl-graphene. Adsorption of PaP1 tail fibers and baseplate to bacterial cell wall resulted in a decrease of ECL signal, since the accumulation of non-conductive bio-complex on electrode surface disrupted the electron transfer. ECL signal dropped linearly with 1.4×10^2–1.4×10^6 CFU/mL concentration of *P. aeruginosa*, with biosensing time of 30 min and very low LOD of 56 colony-forming units per mL. With the help of this biosensor, *P. aeruginosa* was quantified in milk with varying values of recovery from 78.6 to 114.3% [3]. Similarly, phage P100 and magnetic particle composite were established to separate *L. monocytogenes* from food samples. Varied sized magnetic particles (150, 500, and 1000 nm) were used for phage P100 immobilization either physically or chemically. The coupling ratio of composites was investigated, and the capturing efficiency of *L. monocytogenes* was evaluated for each composite. The authors reported that composites developed by physical immobilization of P100 attained a greater efficiency of capture and selectivity toward *L. monocytogenes*. These composites of phage and magnetic particles were further used to selectively isolate *L. monocytogenes* from real sample of food like whole milk and ground beef [46].

3.2 Environmental monitoring

For *E. coli* detection in water, a group of authors established and reported a rise in sensitivity of lateral flow assay based on T7 phage amplification. The assay was founded on phage-based reporter proteins: maltose-binding protein and alkaline phosphatase with 10-folds and 100-folds increased sensitivities, respectively. The increased sensitivity enabled *E. coli* detection of 10^3 CFU/mL in broth while 100 CFU/100 mL of *E. coli* in inoculated river water. Such combination of phage-based diagnosis on paper fluidics offer new platforms to establish innovative detection techniques owing to sensitivity, robustness, and specificity and are personal friendly [47]. Additional improvement in the sensitivity of this method was reported by using fluorescent quantum dots (QDs) for phage labeling. QDs increased the stability and intensity of luminous signal and also enhanced the sensitivity of epifluorescence microscopy and flow cytometry-based detection platforms. The phage head was modified with biotin tagging peptide. The QDs coated with streptavidin were permitted to become bounded to biotinylated bacteriophages. By this approach, detection limit of for *E. coli* was only 20 CFU/mL in water with detection time of 60 min [5].

Similarly, on the basis of phage fluorescent-based detection assays, *Salmonella* in sea water was detected with the help of genetically modified bacteriophages P22 with assay time of 1 h and LOD of 10 CFU/mL [48], while TNT and TNB 1 ng/mL were detected in water with the help of phage display-selected scFv [49]. Likewise, 10^4 CFU/g of *B. anthracis* in soil was detected with the help of Wβ phage involving fluorescence assay [50]. A magnetoelastic biosensor involving JRB7 phage as a bio-recognition element detected 104 spores/mL of *B. anthracis* in water [51], while impedimetric biosensor based on *S. arlettae* specific phage detected 200 CFU/mL of *S. arlettae* in river water [52].

4. Other representative applications

Despite the abovementioned applications of phage-based biosensors, **Table 1** highlights some other representative applications of phage-based biosensors in detection of pathogenic bacteria, food safety, and environmental monitoring.

5. Conclusions and prospects

Without any doubt, environmental monitoring and food safety are the main universal worries that we humans have to oppose and are constantly struggling to take them over. In this chapter, we evidently demonstrated the applications of reported promising platforms of phage-based sensors in the screening of food- and environment-related contaminants. We reviewed demonstrative phage/phage components applied in sensors' development for diagnosis of food pollutants specifically comprising pathogens and toxins. By collaboration with engineers and scientists from multidisciplinary area to design a field applicable sensor and make advancements in phage-based sensors for food safety and environmental monitoring, we expect that this chapter might bring together the technologies related to application of phage-based sensors, in food and environmental safety, and infectious disease diagnostics. In short, applications of phage-based biosensors in the fields of food safety, environmental monitoring, and infectious disease diagnostics are vital.

Acknowledgements

This work was supported by the National Key Research and Development Program of China under Grant 2017YFC1104402 and China Postdoctoral Science Foundation (2016M602291), the initial research fund from CSC, and 3551 Project, Optics Valley of China.

Acronyms and abbreviations

Abs	antibodies
CFU	colony-forming units
E. coli	*Escherichia coli*
EIS	electrochemical impedance spectroscopy
ELISA	enzyme-linked immunosorbent assay
HRP	horseradish peroxidase
ITNAA	isothermal nucleic acid amplification
IUPAC	International Union of Pure and Applied Chemistry
LB	Luria-Bertani broth
LFA	lateral flow assay
LOD	limit of detection
OTA	ochratoxin A
PBS	phosphate-buffered saline
PCR	polymerase chain reaction
PFU	plaque-forming unit
PSA	prostate-specific antigen
QCM	quartz crystal microbalance
QDs	quantum dots
qPCR	quantitative polymerase chain reaction
RBPs	receptor-binding proteins
scFv	single-chain variable fragment
SEB	staphylococcal enterotoxin B
SERS	surface-enhanced Raman spectroscopy
SPR	surface plasmon resonance
TNB	trinitrobenzene
TNT	trinitrotoluene
TZP	tetragonal zirconia polycrystal
BCCP	biotin carboxyl carrier protein
CBM	cellulose-binding module
SOCP	small outer capsid protein
ME	magnetoelastic

Author details

Umer Farooq[1], Muhammad Wajid Ullah[2], Qiaoli Yang[1] and Shenqi Wang[1*]

1 Advanced Biomaterials and Tissue Engineering Centre, Huazhong University of Science and Technology, Wuhan, P. R. China

2 Department of Biomedical Engineering, Huazhong University of Science and Technology, Wuhan, P. R. China

*Address all correspondence to: shenqiwang131@hust.edu.cn

References

[1] Scharff RL. Economic burden from health losses due to foodborne illness in the United States. Journal of Food Protection. 2012;**75**:123-131. DOI: 10.4315/0362-028X.JFP-11-058

[2] Rohde A, Hammerl JA, Boone I, Jansen W, Fohler S, Klein G, et al. Overview of validated alternative methods for the detection of foodborne bacterial pathogens. Trends in Food Science and Technology. 2017;**62**:113-118. DOI: 10.1016/J.TIFS.2017.02.006

[3] Yue H, He Y, Fan E, Wang L, Lu S, Fu Z. Label-free electrochemiluminescent biosensor for rapid and sensitive detection of pseudomonas aeruginosa using phage as highly specific recognition agent. Biosensors & Bioelectronics. 2017;**94**: 429-432. DOI: 10.1016/j.bios.2017.03.033

[4] Park MK, Li S, Chin BA. Detection of *Salmonella typhimurium* grown directly on tomato surface using phage-based magnetoelastic biosensors. Food and Bioprocess Technology. 2013;**6**:682-689. DOI: 10.1007/s11947-011-0708-2

[5] Edgar R, McKinstry M, Hwang J, Oppenheim AB, Fekete RA, Giulian G, et al. High-sensitivity bacterial detection using biotin-tagged phage and quantum-dot nanocomplexes. Proceedings of the National Academy of Sciences. 2006;**103**:4841-4845. DOI: 10.1073/pnas.0601211103

[6] Shabani A, Marquette CA, MandevilleR,LawrenceMF.Magnetically-assisted impedimetric detection of bacteria using phage-modified carbon microarrays. Talanta. 2013;**116**:1047-1053. DOI: 10.1016/j.talanta.2013.07.078

[7] Burnham S, Hu J, Anany H, Brovko L, Deiss F, Derda R, et al. Towards rapid on-site phage-mediated detection of generic *Escherichia coli* in water using luminescent and visual readout. Analytical and Bioanalytical Chemistry. 2014;**406**:5685-5693. DOI: 10.1007/s00216-014-7985-3

[8] Kulpakko J, Kopra K, Hänninen P. Time-resolved fluorescence-based assay for rapid detection of *Escherichia coli*. Analytical Biochemistry. 2015;**470**:1-6. DOI: 10.1016/j.ab.2014.09.002

[9] Martelet A, Lhostis G, Nevers MC, Volland H, Junot C, Becher F, et al. Phage amplification and immunomagnetic separation combined with targeted mass spectrometry for sensitive detection of viable bacteria in complex food matrices. Analytical Chemistry. 2015;**87**:5553-5560. DOI: 10.1021/ac504508a

[10] Karoonuthaisiri N, Charlermroj R, Morton MJ, Oplatowska-Stachowiak M, Grant IR, Elliott CT. Development of a M13 bacteriophage-based SPR detection using salmonella as a case study. Sensors and Actuators B: Chemical. 2014;**190**:214-220. DOI: 10.1016/j.snb.2013.08.068

[11] Kong M, Shin JH, Heu S, Park JK, Ryu S. Lateral flow assay-based bacterial detection using engineered cell wall binding domains of a phage endolysin. Biosensors & Bioelectronics. 2017;**96**:173-177. DOI: 10.1016/j.bios.2017.05.010

[12] Farooq U, Yang Q, Ullah MW, Wang S. Bacterial biosensing: Recent advances in phage-based bioassays and biosensors. Biosensors & Bioelectronics. 2018;**118**:204-216. DOI: 10.1016/j.bios.2018.07.058

[13] Sharp NJ, Vandamm JP, Molineux IJ, Schofield DA. Rapid detection of bacillus anthracis in complex food matrices using phage-mediated bioluminescence. Journal of Food Protection.

2015;**78**:963-968. DOI: 10.4315/0362-028X.JFP-14-534

[14] Singh A, Poshtiban S, Evoy S. Recent advances in bacteriophage based biosensors for food-borne pathogen detection. Sensors. 2013;**13**:1763-1786. DOI: 10.3390/s130201763

[15] Arya SK, Singh A, Naidoo R, Wu P, McDermott MT, Evoy S. Chemically immobilized T4-bacteriophage for specific *Escherichia coli* detection using surface plasmon resonance. The Analyst. 2011;**136**:486-492. DOI: 10.1039/C0AN00697A

[16] Balasubramanian S, Sorokulova IB, Vodyanoy VJ, Simonian AL. Lytic phage as a specific and selective probe for detection of *Staphylococcus aureus*-A surface plasmon resonance spectroscopic study. Biosensors & Bioelectronics. 2007;**22**:948-955. DOI: 10.1016/j.bios.2006.04.003

[17] Tawil N, Sacher E, Mandeville R, Meunier M. Surface plasmon resonance detection of *E. coli* and methicillin-resistant *S. aureus* using bacteriophages. Biosensors & Bioelectronics. 2012;**37**: 24-29. DOI: 10.1016/j.bios.2012.04.048

[18] Singh A, Arya SK, Glass N, Hanifi-Moghaddam P, Naidoo R, Szymanski CM, et al. Bacteriophage tailspike proteins as molecular probes for sensitive and selective bacterial detection. Biosensors & Bioelectronics. 2010;**26**:131-138. DOI: 10.1016/j.bios.2010.05.024

[19] Blasco R, Murphy MJ, Sanders MF, Squirrell DJ. Specific assays for bacteria using phage mediated release of adenylate kinase. Journal of Applied Microbiology. 1998;**84**:661-666. DOI: 10.1046/j.1365-2672.1998.00393.x

[20] Wu Y, Brovko L, Griffiths MW. Influence of phage population on the phage-mediated bioluminescent adenylate kinase (AK) assay for detection of bacteria. Letters in Applied Microbiology. 2001;**33**:311-315. DOI: 10.1046/j.1472-765X.2001.01002.x

[21] Peltomaa R, López-Perolio I, Benito-Peña E, Barderas R, Moreno-Bondi MC. Application of bacteriophages in sensor development. Analytical and Bioanalytical Chemistry. 2016;**408**:1805-1828. DOI: 10.1007/s00216-015-9087-2

[22] Srivastava SK, Ben Hamo H, Kushmaro A, Marks RS, Grüner C, Rauschenbach B, et al. Highly sensitive and specific detection of *E. coli* by a SERS nanobiosensor chip utilizing metallic nanosculptured thin films. Analyst. 2015;**140**:3201-3209. DOI: 10.1039/C5AN00209E

[23] Goodridge L, Chen J, Griffiths M. The use of a fluorescent bacteriophage assay for detection of *Escherichia coli* O157:H7 in inoculated ground beef and raw milk. International Journal of Food Microbiology. 1999;**47**:43-50. DOI: 10.1016/S0168-1605(99)00010-0

[24] Goodridge L, Chen J, Griffiths M. Development and characterization of a fluorescent-bacteriophage assay for detection of *Escherichia coli* O157:H7. Applied and Environmental Microbiology. 1999;**65**:1397-1404

[25] Yim PB, Clarke ML, McKinstry M, De Paoli Lacerda SH, Pease LF, Dobrovolskaia MA, et al. Quantitative characterization of quantum dot-labeled lambda phage for *Escherichia coli* detection. Biotechnology and Bioengineering. 2009;**104**:1059-1067. DOI: 10.1002/bit.22488

[26] Goldman ER, Pazirandeh MP, Matthew Mauro J, King KD, Frey JC, Anderson GP. Phage-displayed peptides as biosensor reagents. Journal of Molecular Recognition. 2000;**13**(6):382-387. DOI: 10.1002/1099-1352(200011/12)13:6<382:AID-JMR511>3.0.CO;2-W

[27] Rowe CA, Tender LM, Feldstein MJ, Golden JP, Scruggs SB, MacCraith BD, et al. Array biosensor for simultaneous identification of bacterial, viral, and protein analytes. Analytical Chemistry. 1999;**71**(17):3846-3852. DOI: 10.1021/ac981425v

[28] Wu L, Song Y, Luan T, Ma L, Su L, Wang S, et al. Specific detection of live *Escherichia coli* O157: H7 using tetracysteine-tagged PP01 bacteriophage. Biosensors & Bioelectronics. 2016;**86**:102-108. DOI: 10.1016/j.bios.2016.06.041

[29] Smartt AE, Ripp S. Bacteriophage reporter technology for sensing and detecting microbial targets. Analytical and Bioanalytical Chemistry. 2011;**400**:991-1007. DOI: 10.1007/s00216-010-4561-3

[30] Irwin P, Gehring A, Shu-I TU, Brewster J, Fanelli J, Ehrenfeld E. Minimum detectable level of salmonellae using a binomial-ice based bacterial ice nucleation detection assay (BIND). Journal of AOAC International. 2000;**83**(5):1087-1095

[31] Schmelcher M, Loessner MJ. Application of bacteriophages for detection of foodborne pathogens. Bacteriophage. 2014;**4**:128-137. DOI: 10.4161/bact.28137

[32] Rossotti MA, Carlomagno M, González-Techera A, Hammock BD, Last J, González-Sapienza G. Phage anti-immunocomplex assay for clomazone: Two-site recognition increasing assay specificity and facilitating adaptation into an on-site format. Analytical Chemistry. 2010;**82**(21):8838-8843. DOI: 10.1021/ac101476f

[33] Oh JW, Chung WJ, Heo K, Jin HE, Lee BY, Wang E, et al. Biomimetic virus-based colourimetric sensors. Nature Communications. 2014;**5**:3043. DOI: 10.1038/ncomms4043

[34] Arugula MA, Simonian A. Novel trends in affinity biosensors: Current challenges and perspectives. Measurement Science and Technology. 2014;**25**(3):032001. DOI: 10.1088/0957-0233/25/3/032001

[35] Olsen EV, Sorokulova IB, Petrenko VA, Chen IH, Barbaree JM, Vodyanoy VJ. Affinity-selected filamentous bacteriophage as a probe for acoustic wave biodetectors of *Salmonella typhimurium*. Biosensors & Bioelectronics. 2006;**21**:1434-1442. DOI: 10.1016/j.bios.2005.06.004

[36] Petrenko VA, Sorokulova IB. Detection of biological threats. A challenge for directed molecular evolution. Journal of Microbiological Methods. 2004;**58**(2):147-168. DOI: 10.1016/j.mimet.2004.04.004

[37] Park MK, Park JW, Wikle HC, Chin BA. Evaluation of phage-based magnetoelastic biosensors for direct detection of *Salmonella typhimurium* on spinach leaves. Sensors and Actuators B: Chemical. 2013;**176**:1134-1140. DOI: 10.1016/j.snb.2012.10.084

[38] Lakshmanan RS, Guntupalli R, Hu J, Petrenko VA, Barbaree JM, Chin BA. Detection of *Salmonella typhimurium* in fat free milk using a phage immobilized magnetoelastic sensor. Sensors and Actuators B: Chemical. 2007;**126**(2):544-550. DOI: 10.1016/j.snb.2007.04.003

[39] Guntupalli R, Hu J, Lakshmanan RS, Huang TS, Barbaree JM, Chin BA. A magnetoelastic resonance biosensor immobilized with polyclonal antibody for the detection of *Salmonella typhimurium*. Biosensors & Bioelectronics. 2007;**22**(7):1474-1479. DOI: 10.1016/j.bios.2006.06.037

[40] Hiremath N, Guntupalli R, Vodyanoy V, Chin BA, Park MK. Detection of methicillin-resistant *Staphylococcus aureus* using novel lytic

phage-based magnetoelastic biosensors. Sensors and Actuators B: Chemical. 2015;**210**:129-136. DOI: 10.1016/j.snb.2014.12.083

[41] Li Y, Xie G, Qiu J, Zhou D, Gou D, Tao Y, et al. A new biosensor based on the recognition of phages and the signal amplification of organic-inorganic hybrid nanoflowers for discriminating and quantitating live pathogenic bacteria in urine. Sensors and Actuators B: Chemical. 2018;**258**:803-812. DOI: 10.1016/j.snb.2017.11.155

[42] Neufeld T, Schwartz-Mittelmann A, Biran D, Ron EZ, Rishpon J. Combined phage typing and amperometric detection of released enzymatic activity for the specific identification and quantification of bacteria. Analytical Chemistry. 2003;**75**:580-585. DOI: 10.1021/ac026083e

[43] Yemini M, Levi Y, Yagil E, Rishpon J. Specific electrochemical phage sensing for *Bacillus cereus* and *Mycobacterium smegmatis*. Bioelectrochemistry. 2007;**70**:180-184. DOI: 10.1016/j.bioelechem.2006.03.014

[44] Shabani A, Zourob M, Allain B, Marquette CA, Lawrence MF, Mandeville R. Bacteriophage-modified microarrays for the direct impedimetric detection of bacteria. Analytical Chemistry. 2008;**80**:9475-9482. DOI: 10.1021/ac801607w

[45] Chen IH, Horikawa S, Bryant K, Riggs R, Chin BA, Barbaree JM. Bacterial assessment of phage magnetoelastic sensors for *Salmonella enterica* typhimurium detection in chicken meat. Food Control. 2017;**71**:273-278. DOI: 10.1016/j.foodcont.2016.07.003

[46] Zhou Y, Ramasamy RP. Isolation and separation of listeria monocytogenes using bacteriophage P100-modified magnetic particles. Colloids and Surfaces. B, Biointerfaces. 2018;**175**:421-427. DOI: 10.1016/j.colsurfb.2018.12.007

[47] Alcaine SD, Law K, Ho S, Kinchla AJ, Sela DA, Nugen SR. Bioengineering bacteriophages to enhance the sensitivity of phage amplification-based paper fluidic detection of bacteria. Biosensors & Bioelectronics. 2016;**82**:14-19. DOI: 10.1016/j.bios.2016.03.047

[48] Vinay M, Franche N, Grégori G, Fantino JR, Pouillot F, Ansaldi M. Phage-based fluorescent biosensor prototypes to specifically detect enteric bacteria such as *E. coli* and *Salmonella enterica* typhimurium. PLoS One. 2015;**10**:e0131466. DOI: 10.1371/journal.pone.0131466

[49] Goldman ER, Hayhurst A, Lingerfelt BM, Iverson BL, Georgiou G, Anderson GP. 2,4,6-Trinitrotoluene detection using recombinant antibodies. Journal of Environmental Monitoring. 2003;**5**(3):380-383. DOI: 10.1039/b302012f

[50] Sharp NJ, Molineux IJ, Page MA, Schofield DA. Rapid detection of viable *Bacillus anthracis* spores in environmental samples by using engineered reporter phages. Applied and Environmental Microbiology. 2016;**82**:2380-2387. DOI: 10.1128/AEM.03772-15

[51] Fu L, Li S, Zhang K, Chen IH, Barbaree JM, Zhang A, et al. Detection of *Bacillus anthracis* spores using phage-immobilized magnetostrictive milli/micro cantilevers. IEEE Sensors Journal. 2011;**11**(8):1684-1691. DOI: 10.1109/JSEN.2010.2095002

[52] Bhardwaj N, Bhardwaj SK, Mehta J, Mohanta GC, Deep A. Bacteriophage immobilized graphene electrodes for impedimetric sensing of bacteria (*Staphylococcus arlettae*). Analytical Biochemistry. 2016;**505**:18-25. DOI: 10.1016/j.ab.2016.04.008

[53] Nanduri V, Sorokulova IB, Samoylov AM, Simonian AL, Petrenko VA,

Vodyanoy V. Phage as a molecular recognition element in biosensors immobilized by physical adsorption. Biosensors & Bioelectronics. 2007;**22**(6):986-992. DOI: 10.1016/j.bios.2006.03.025

[54] Erzinger GS, Schmoeller F, Pinto LH, Américo L, Hemmersbach R, Hauslage J, et al. Bioluminescence systems in environmental biosensors. In: Bioassays Adv. Methods Appl. Netherland: Elsevier; 2017. DOI: 10.1016/B978-0-12-811861-0.00012-7

[55] Schofield DA, Wray DJ, Molineux IJ. Isolation and development of bioluminescent reporter phages for bacterial dysentery. European Journal of Clinical Microbiology & Infectious Diseases. 2015;**34**(2):395-403. DOI: 10.1007/s10096-014-2246-0

[56] Cox CR, Jensen KR, Mondesire RR, Voorhees KJ. Rapid detection of *Bacillus anthracis* by γ phage amplification and lateral flow immunochromatography. Journal of Microbiological Methods. 2015;**118**:51-56. DOI: 10.1016/j.mimet.2015.08.011

[57] Laube T, Cortés P, Llagostera M, Alegret S, Pividori MI. Phagomagnetic immunoassay for the rapid detection of *Salmonella*. Applied Microbiology and Biotechnology. 2014;**98**:1795-1805. DOI: 10.1007/s00253-013-5434-4

[58] Yan C, Zhang Y, Yang H, Yu J, Wei H. Combining phagomagnetic separation with immunoassay for specific, fast and sensitive detection of *Staphylococcus aureus*. Talanta. 2017;**170**:291-297. DOI: 10.1016/j.talanta.2017.04.007

[59] Zhang Y, Yan C, Yang H, Yu J, Wei H. Rapid and selective detection of *E. coli* O157:H7 combining phagomagnetic separation with enzymatic colorimetry. Food Chemistry. 2017;**234**:332-338. DOI: 10.1016/j.foodchem.2017.05.013

[60] Hoang HA, Dien LT. Rapid and simple colorimetric detection of *Escherichia coli* O157:H7 in apple juice using a novel recombinant bacteriophage-based method. Biocontrol Science. 2015;**20**:99-103. DOI: 10.4265/bio.20.99

[61] Goldman ER, Pazirandeh MP, Charles PT, Balighian ED, Anderson GP. Selection of phage displayed peptides for the detection of 2,4,6-trinitrotoluene in seawater. Analytica Chimica Acta. 2002;**457**(1):13-19. DOI: 10.1016/S0003-2670(01)01246-6

[62] Chen J, Alcaine SD, Jiang Z, Rotello VM, Nugen SR. Detection of *Escherichia coli* in drinking water using T7 bacteriophage-conjugated magnetic probe. Analytical Chemistry. 2015;**87**:8977-8984. DOI: 10.1021/acs.analchem.5b02175

[63] Lee HE, Lee HK, Chang H, Ahn HY, Erdene N, Lee HY, et al. Virus templated gold nanocube chain for SERS nanoprobe. Small. 2014;**10**(15):3007-3011. DOI: 10.1002/smll.201400527

[64] Lentini G, Franco D, Fazio E, De Plano LM, Trusso S, Carnazza S, et al. Rapid detection of *Pseudomonas aeruginosa* by phage-capture system coupled with micro-Raman spectroscopy. Vibrational Spectroscopy. 2016;**86**:1-7. DOI: 10.1016/j.vibspec.2016.05.003

[65] Stambach NR, Carr SA, Cox CR, Voorhees KJ. Rapid detection of listeria by bacteriophage amplification and SERS-lateral flow immunochromatography. Viruses. 2015;7(12):6631-6641. DOI: 10.3390/v7122962

[66] Li S, Li Y, Chen H, Horikawa S, Shen W, Simonian A, et al. Direct detection of *Salmonella typhimurium* on fresh produce using phage-based magnetoelastic biosensors. Biosensors & Bioelectronics.

2010;**26**:1313-1319. DOI: 10.1016/j.bios.2010.07.029

[67] Shen W, Lakshmanan RS, MathisonLC, PetrenkoVA, ChinBA. Phage coated magnetoelastic micro-biosensors for real-time detection of *Bacillus anthracis* spores. Sensors and Actuators B: Chemical. 2009;**137**:501-506. DOI: 10.1016/j.snb.2009.01.027

[68] Mack JD, Yehualaeshet T, Park MK, Tameru B, Samuel T, Chin BA. Phage-based biosensor and optimization of surface blocking agents to detect *Salmonella typhimurium* on Romaine lettuce. Journal of Food Safety. 2017;**37**(2):e12299. DOI: 10.1111/jfs.12299

[69] Chai Y, Horikawa S, Li S, Wikle HC, Chin BA. A surface-scanning coil detector for real-time, in-situ detection of bacteria on fresh food surfaces. Biosensors & Bioelectronics. 2013;**50**:311-317. DOI: 10.1016/j.bios.2013.06.056

[70] Kretzer JW, Lehmann R, Schmelcher M, Banz M, Kim KP, Korn C, et al. Use of high-affinity cell wall-binding domains of bacteriophage endolysins for immobilization and separation of bacterial cells. Applied and Environmental Microbiology. 2007;**73**(6):1992-2000. DOI: 10.1128/AEM.02402-06

[71] Neufeld T, Mittelman AS, Buchner V, Rishpon J. Electrochemical phagemid assay for the specific detection of bacteria using *Escherichia coli* TG-1 and the M13KO7 phagemid in a model system. Analytical Chemistry. 2005;**77**:652-657. DOI: 10.1021/ac0488053

[72] Lee SH, Onuki M, Satoh H, Mino T. Isolation, characterization of bacteriophages specific to *Microlunatus phosphovorus* and their application for rapid host detection. Letters in Applied Microbiology. 2006;**42**:259-264. DOI: 10.1111/j.1472-765X.2006.01840.x

[73] Zhou Y, Marar A, Kner P, Ramasamy RP. Charge-directed immobilization of bacteriophage on nanostructured electrode for whole-cell electrochemical biosensors. Analytical Chemistry. 2017;**89**:5734-5741. DOI: 10.1021/acs.analchem.6b03751

[74] Dadarwal R, Namvar A, Thomas DF, Hall JC, Warriner K. Organic conducting polymer electrode based sensors for detection of *Salmonella* infecting bacteriophages. Materials Science and Engineering: C. 2009;**29**(3):761-765. DOI: 10.1016/j.msec.2008.07.010

[75] Shabani A, Marquette CA, Mandeville R, Lawrence MF. Carbon microarrays for the direct impedimetric detection of *Bacillus anthracis* using gamma phages as probes. The Analyst. 2013;**138**:1434. DOI: 10.1039/c3an36830k

[76] Tlili C, Sokullu E, Safavieh M, Tolba M, Ahmed MU, Zourob M. Bacteria screening, viability, and confirmation assays using bacteriophage-impedimetric/loop-mediated isothermal amplification dual-response biosensors. Analytical Chemistry. 2013;**85**: 4893-4901. DOI: 10.1021/ac302699x

[77] Tolba M, Ahmed MU, Tlili C, Eichenseher F, Loessner MJ, Zourob M. A bacteriophage endolysin-based electrochemical impedance biosensor for the rapid detection of Listeria cells. The Analyst. 2012;**137**:5749. DOI: 10.1039/c2an35988j